MUSHROOMS
of Western North America

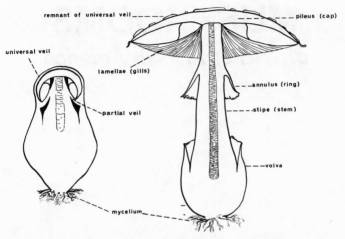

Diagram showing major macroscopic parts of *Amanita* sp. in button stage (left) and at maturity (right). Other genera may have only an annulus or a volva, or they may lack both of these structures.

California Natural History Guides: 42

MUSHROOMS

of Western North America

Robert T. Orr and Dorothy B. Orr

Drawings by

Jacqueline Schonewald and Paul Vergeer

Color illustrations by the authors

UNIVERSITY OF CALIFORNIA PRESS

Berkeley • Los Angeles • London

University of California Press
Berkeley and Los Angeles, California
University of California Press, Ltd.
London, England

©1979 by The Regents of the University of California
Library of Congress Catalog Card Number: 77-93468
Printed in the United States of America

1 2 3 4 5 6 7 8 9

CONTENTS

1. INTRODUCTION

2. ASCOMYCETES/Sac Fungi

3. BASIDIOMYCETES/Club Fungi: *Nongilled Fungi*

4. BASIDIOMYCETES/Club Fungi: *Gilled Fungi* (Agarics)

1. INTRODUCTION

This field guide is an outgrowth of two previous regional booklets by the authors published in the California Natural History Guide series. The first, *Mushrooms and Other Common Fungi of the San Francisco Bay Region*, covered those fleshy and woody species most frequently encountered in coastal northern California. The second, *Mushrooms and Other Common Fungi of Southern California*, included species of southern coastal and montane areas as well as desert regions of the state.

The present work encompasses the area extending along the Pacific Coast from southern California east through the Rocky Mountains to the western edge of the Great Plains, and north to British Columbia, western Alberta, the Yukon Territory, and Alaska. No single book could possibly describe the thousands of kinds of fleshy fungi that occur in such an extensive area. However, we hope that the user of this guide will soon learn to distinguish the major groups and many of the genera, as well as the more common species. Many of the mushrooms included occur widely over this region, while others are rather restricted; in some instances their presence is correlated with that of certain forest trees.

Obviously many factors, such as elevation, rainfall, and temperature, affect the distribution of mushrooms in this widespread and diverse area. There is a range of habitat from deserts (where stalked puffballs seem to be about the only fungi capable of surviving) to lush countrysides and conifer or hardwood forests. The elevation may be below sea level in Death Valley, California, to above timberline in some mountain ranges.

1

We have indicated the general distribution of the species to the best of our knowledge, but you can reasonably expect to find any of these species, whether recorded or not in your collecting area, if they are growing in a similar habitat, at approximately the same elevation, and with appropriate weather conditions.

Relatively little has been written about the Alaskan and western Canadian mycoflora, but those species known to occur in the Pacific Northwest are very likely to be found in the adjacent provinces also, except their mushroom season may come a bit earlier with the summer rains.

It is frustrating to find a specimen that fails to fit a description in your reference book, and this can happen frequently. Some genera contain hundreds of species (for instance, *Cortinarius* is presently estimated to have over 600 species). Also, new species and even genera are being found all the time, and often there is a lapse of many months before the description is published. The serious student who wishes to carry identification further must consult special monographs.

To aid you in using the species descriptions in this book, the frontispiece illustrates the major macroscopic parts of mushrooms. A Glossary is included at the end of the book for definitions used routinely in this and most other mushroom books. You will soon become familiar with these terms.

The study of mushrooms has become increasingly popular in the United States and Canada during the past quarter-century, although it has long been a way of life in European countries for freely obtaining delicious additions to the table. In the West, where the mushroom flora is prolific in good years, there are now a number of mycological societies, which are listed at the end of the book. These groups meet regularly to attend lectures and participate in field trips led by experts. Many societies also put on annual Fungus Fairs for the general public.

THE CHARACTERISTICS OF FUNGI

Fungi, like bacteria and slime molds, lack chlorophyll, the green coloring matter which enables most plants to manufacture their food from carbon dioxide and water in the presence of sunlight. Since they are unable to synthesize food materials, fungi must use food that has already been produced, and they accomplish this by subsisting on other live organisms or on dead organic material. The former method is called *parasitism*; the latter, *saprophytism*.

A few kinds of fungi parasitize man and other animals and are the cause of diseases such as athlete's foot and ringworm. Some are parasitic on other plants and often cause great economic loss. Many fruit and ornamental trees are killed each year by one of our most common species, the "honey mushroom," *Armillariella mellea*. A few even parasitize other fungi.

Fortunately, the majority of the many thousands of species of fungi in the world are saprophytes. Since all living organisms eventually die, there is a wealth of food for those that live on dead organic matter. This function is absolutely essential to our earth, because without it we would be overwhelmed with dead trees and other plants. Sometimes, as in stumps and logs, the food survives for many years before it is completely utilized. In the process of bringing about the decomposition of plant and animal material, saprophytic fungi, like many bacteria, enrich the soil and provide food for higher plants. Fungi that live on wood are called *lignicolous*; those that live on the ground are *terrestrial*. There are also many aquatic species which include important pathogens of higher plants that may also parasitize insects, fish, domestic animals, and man.

A few fungi are unicellular, such as the yeasts, but most of them are made up of filamentous, branching aggregations of cells called *hyphae*. The mass of hyphae comprising a fungus plant is known as a *mycelium*. Good places to look for

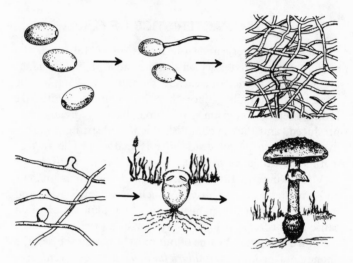

FIG. 1. Life cycle of a gilled mushroom from spore, hypha, mycelium, clamp connection, button stage, to mature fruiting body.

mycelia are under the loose bark of rotting trees or under damp boards on the ground. From the mycelium arises the fruiting body, called the *sporophore* or *carpophore*; this is the spore-producing organ (see Fig. 1). The fruiting bodies of fleshy fungi have commonly been known as *mushrooms* and *toadstools*—these are vague distinctions supposed to indicate edible and poisonous species of fungi, but it is now standard to call them all mushrooms and indicate whether they are edible, poisonous, or merely unpalatable because of poor flavor, coarse texture, or lack of substance when cooked (see discussion on edibility below, p. 16; where known, this information is listed in species descriptions).

Fleshy fungi are technically divided into two main classes: the *Ascomycetes*, or sac fungi, and the *Basidiomycetes*, or club fungi. The Ascomycetes are fungi in which the spores are borne in sacs known as *asci*. Ascomycetes that will be considered here are the cup fungi, morels, and saddle fungi, but the class also includes yeasts, molds, and powdery mil-

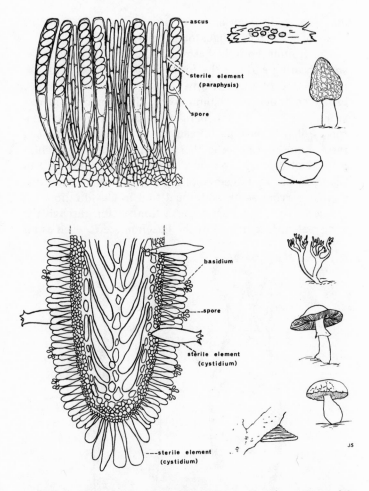

FIG. 2. Microscopic sections through spore-bearing or hymenial surface of a cup fungus, an Ascomycete (above), and the gill of an agaric, a Basidiomycete (below).

dews. The class Basidiomycetes contains most of the fleshy and woody fungi, as well as the rusts and smuts. In the reproductive cycle, their spores are produced on clublike structures called *basidia*; these basidia may be on the gills of an *agaric* (gilled mushroom), inside a puffball, or in the tubes of a bracket fungus or a bolete (see Fig. 2).

The fruiting body in fleshy fungi of both classes has a spore-bearing surface, the *hymenium*. In a razor-thin section of one of the Ascomycetes, such as a cup fungus, examined under the medium power of a compound microscope, one can see the parallel series of asci containing spores, if they have not already been discharged. Between these spore-bearing cells there are numerous sterile filaments called *paraphyses*. In different species these vary in shape, as well as in their color reaction to certain chemicals, thereby serving as an additional tool in classification.

Microscopic examination of a thin section through the hymenial surface of one of the Basidiomycetes, such as an

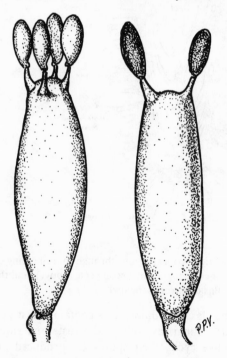

FIG. 3. Two- and four-spored basidia.

agaric, will shown the club-shaped basidia with 2, 3, or 4 *spicules* called *sterigmata* at the tip (see Fig. 3); the spores are attached to these tiny appendages. Many *basidioles*—smaller immature basidia lacking sterigmata—are interspersed between the mature basidia. In some Basidiomycetes, larger sterile cells of a different shape, called *cystidia*, are also present on parts of the hymenium. Those occurring on the sides of a gill are called *pleurocystidia*; those on the edge of a gill are *cheilocystidia*. Cystidial cells also occur on the *pileus* (cap) as well as on the *stipe* (stem) of a fruiting body.

All measurements for species in this guide are given in the metric system. Gross measurements are in centimeters (cm) and microscopic measurements in microns, represented by the symbol μ (mu). A centimeter is one one-hundredth of a meter, and a micron is one-millionth of a meter.

SEASONAL OCCURRENCE AND HABITAT

Climate is a major factor affecting the seasonal occurrence of fungi. In the eastern United States, where summer rains are frequent, fungi fruit most abundantly in late summer and early autumn. With the arrival of cold weather in late autumn, growth ceases for the year.

In western North America, the great climatic diversity makes any generalization on the overall fungus season impossible. There is never a month when fungi cannot be found somewhere in the West, but the governing factors are moisture and temperature. In midsummer in the high Sierra Nevada, the Cascades, and the Rocky Mountains, there are fungi that appear along the edge of snowbanks near the timberline. Summer rains in the Cascades may cause abundant fruiting of fungi in August. In the mountains of Arizona, the fungus season is dependent on summer rains in July and August. Late summer rains in the northern Rocky Mountains make September the prime season there, while October is usually the autumn fungus season in the Sierra Nevada and the Pacific Northwest. Farther south along the coast the season is progressively later. In central California fungi are rarely found in quantity before late November of early December, and in southern California. January, February, and March are the favored months. With the arrival of spring and the melting of snow in the higher mountains of the West, fungi once again begin to appear in May and June at mid-elevations.

Although the time of the fruiting season varies greatly from region to region, it is usually fairly constant in any given area. For example, fungi are very abundant in the coastal forests of Oregon from October through December. By February, however, even with an abundance of moisture, the fruiting season is over and scarcely a specimen can be found. The behavior of fungi is no different from that of most plants: there is a time for reproduction and a time for rest.

The seasonal abundance or scarcity of fungi in any region is naturally influenced by rainfall and temperature. In years of drought, the onset of the fungus season may be delayed and the period of fruiting greatly shortened; extreme drought may almost eliminate it. A successful fungus season may be abruptly terminated by unseasonable cold and early snowfall. It will also be adversely affected by excessive rainfall.

While a few kinds of fleshy fungi occur in open grassland, usually in pastures well fertilized by domestic stock, the great majority are forest inhabitants, because the food material they require for growth is available there. This food consists mostly of accumulated duff, composed of leaves and other decaying vegetation on the forest floor, as well as rotting logs, stumps, or living trees. Among the lignicolous fungi, some are parasitic on living trees and others are saprophytic, living on the remains of dead trees. Some species, like members of the *Armillariella mellea* complex, may have either dead or living trees as host.

MYCORRHIZAL ASSOCIATIONS

Some fungi have an intimate association with certain trees. This association occurs between the mycelium and the tiny terminal rootlets of the larger plants; the mycelial threads may be wrapped around the rootlets or they may penetrate them, but they do not invade the larger roots and become parasitic. This close relationship between the fungus and the forest tree seems to be mutually beneficial, or *symbiotic*, and is referred to as a *mycorrhiza*. Thus some fungi are found only in association with oaks or with conifers; some are even more restricted and occur only under two- or three-needle pines. *Suillus pungens* of coastal California seems to be associated with Monterey Pine (*Pinus radiata*), a tree of very limited natural distribution but one which has been successfully introduced into many other parts of the world as an ornamental as well as a timber species. Several years ago *S. pungens* was observed growing under introduced *P. radiata* at the airport in Bogota, Colombia; neither tree nor fungus is native to that country, but one had accompanied the other. Another bolete, *Suillus grevillei*, occurs only under or near larch (*Larix* spp.). *Fuscoboletinus ochraceoroseus* seems to be even more restricted, occurring only with Western Larch (*L. occidentalis*). A knowledge of the mycorrhizal association of many kinds of fungi with certain forest trees is an important aid to the mushroom hunter.

COLLECTING FUNGI

Mushroom collecting can be both gastronomically and scientifically productive, as well as exciting and good out-of-doors exercise. It should be done properly, however. The equipment required is minimal: a basket or some other easily carried container, a supply of waxed paper, a knife, and a trowel. Specimens should be wrapped in waxed paper, with the ends of the paper twisted to conserve moisture and keep the mushrooms as fresh as possible. (Don't use plastic bags for carrying mushrooms, as they cause rapid deterioration of any fresh material. Waxed paper does not produce an airtight seal or cause sweating as do plastic bags.) Small or very delicate mushrooms should be surrounded by moss or grass to keep them from being crushed; small boxes may help to protect a particular rarity. Avoid putting large or heavy specimens on top of smaller, more fragile ones.

When collecting fungi for study, it is important to secure the entire fruiting body. The cap, gills, tubes, or stipe of some species (especially among the boletes) may change color when cut. To determine if a specimen will react this way, make a small incision with a knife in the appropriate parts. In the genus *Lactarius*, for example, the color of the juice, as well as changes that may occur on exposure to air, are important characters in field identification. Keep specimens cool while in the field, and refrigerate them, after returning, until you are ready to work on them; do not, however, freeze them. Mushrooms identified with certainty and picked for eating should be kept separate, with as much dirt removed as possible.

Those who wish to make a serious study of fungi should keep a notebook and obtain spore prints of each species collected. For each specimen, record your serial number; its name, if known; the specific locality where collected; its habitat; the date; and an accurate description of the fruiting body. Spore prints of fleshy fungi can be obtained by removing the cap from the stipe, if present, and placing the spore-bearing surface down on a piece of white paper for several hours until a deposit is visible. If small pieces of white paper are included in the collecting basket, a spore print may be started in the field and will often be complete by the time you reach home. Even white spores can be seen on white paper. The color of the spores is important in identification.

In recording the description of a mushroom, the characters to be noted include size, shape, color, texture, odor, and "feel" of the specimen. It may be dry, moist, slimy, or viscid (sticky). The *pileus* (cap) may be smooth, scaly, striate, convex, concave, conic, umbonate, plane, funnel-shaped, etc. (see Fig. 4). In an *agaric* (gilled mushroom), the *lamellae* (gills) may be free from the *stipe* (stem) or attached in varying degrees; they may be shallow or deep, close together or distant; there may be additional small gills called *lamellulae* that alternate with the lamellae. In pore fungi, the tubes and their shape should be noted. In coral fungi, the degree of branching and the shape of the tips of the branches are important for identification. For *stipitate* (stemmed) fungi, record the measurements of the stipe and its shape (see Fig. 5), and note the presence or absence of an *annulus* (a membranous ring on the stipe) or a *volva* (a cuplike sheath at the base of the stipe). Always record any pronounced odor; some fungi have distinct odors similar to anise, cedar, almond, chlorine, grain, radish, apricot, etc. For mushrooms positively identified as edible, record the taste: acrid, peppery, sweet, mild, etc. The color of the mycelium is sometimes significant and should be noted.

If you want to keep a reference collection at home, you must dry your specimens. This can be done simply by replacing the tray of a small TV table with quarter-inch hardware cloth or screen and placing it over a furnace outlet. The warm, dry, circulating air will dry all but the largest specimens rapidly, without burning or otherwise injuring the cellular material. Large specimens may be cut in half or even sectioned. Small portable dryers can be made by constructing a frame with screen trays over an electric hotplate. Another practical method for drying small fungi, like members of the genus *Mycena*, is to use silica gel, which is readily obtainable; the blue crystals absorb the moisture from the fungi. After repeated use, the crystals change to pink and must be heated to remove the water and return them to the dry blue state.

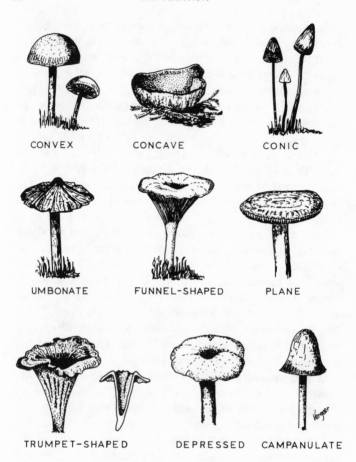

CONVEX   CONCAVE   CONIC

UMBONATE   FUNNEL-SHAPED   PLANE

TRUMPET-SHAPED   DEPRESSED   CAMPANULATE

FIG. 4. Typical pileus (cap) shapes.

Dry specimens should be stored in paper boxes; the type of box used will depend upon the ingenuity and budget of the collector. Each box should have a label bearing the name and field number of the specimen, the locality, date collected, and your name. If detailed notes are kept, the field number will be an adequate reference to them. If a spore print has been made, it should be kept with the specimen. A

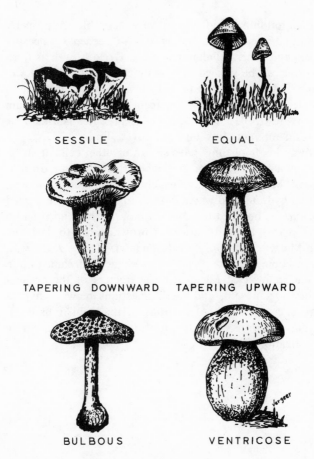

SESSILE   EQUAL

TAPERING DOWNWARD   TAPERING UPWARD

BULBOUS   VENTRICOSE

FIG. 5. Typical stipe (stem) shapes.

suitable insect deterrent, such as napthalene crystals, should be placed in each box to prevent infestation by carpet beetles and weevils.

For the serious student of mycology, a compound microscope is an essential tool. Microscopic characters are even more important in the study of mycology than macroscopic characters. Medium and high dry objectives are essential, and an oil-immersion lens is very desirable. A properly calibrated occular micrometer is also needed to measure spores and other cellular structures. Dried as well as fresh material can be studied microscopically. A razor-thin section of a dried fungus can rapidly be revived by placing it on a slide with a drop of 2.5-percent solution of potassium hydroxide (KOH). For fresh material, water is a suitable medium for mounting under a cover slip.

Many kinds of fungi have spores or other cellular structures that turn bluish or gray when treated with a drop or two of Melzer's solution, which is composed of iodine, chloral hydrate, and water. Spores that so react are termed *amyloid* (as opposed to non-amyloid, in which no reaction takes place). Some tissues turn brownish-red to dark red with Melzer's solution, a condition referred to as *dextrinoid*. Microscopic characters such as the shape or ornamentation of the spores (see Fig. 6), the characters of the cystidia if they are present, and the connections between hyphal strands, as well as the reactions of the cellular tissue to various solutions, are all important in making accurate identifications.

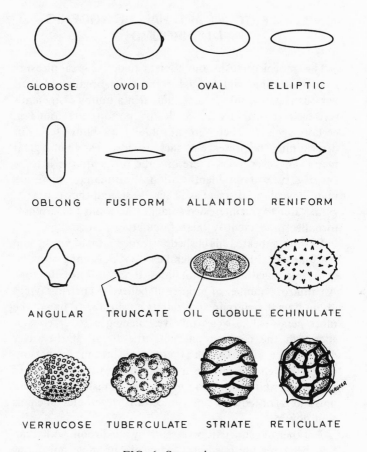

FIG. 6. Spore shapes.

EDIBLE, TOXIC, AND HALLUCINOGENIC
MUSHROOMS

The informed mushroom hunter can derive great pleasure from collecting wild fungi for the table because of their various flavors and textures, but such a hobby also entails serious responsibility, since deadly poisons are found in certain species. There are absolutely no home tests for determining the presence of these toxins, in spite of a great many old wives' tales; therefore, the first requisite is to be entirely sure about identification. Fortunately, there are many mycological societies in the West, and they can be of great value to the interested student who wants to learn how to collect and identify these fascinating plants.

Despite repeated published warnings, each mushroom season inevitably brings several cases of poisoning or hallu-cinogenic reactions, even fatalities, to incautious collectors. An amazing number of people still believe in ancient sayings about how to determine poisonous mushrooms—one of the most persistent fallacies involves placing a silver coin or spoon in the cooking pot with the fungi. If the silver darkens, it is taken to mean that the mushrooms are poison-ous, although this change merely indicates the presence of sulphides, which may be found in many mushrooms (either poisonous or edible) and in certain vegetables such as cauliflower.

Another old superstition is that any mushroom with a cap that peels readily is edible. Many of our most poisonous Amanitas would pass this "test" easily! Some people also continue to rely on the theory that any kind of fungus which is eaten by animals is automatically safe for human con-sumption; but this is not true, as some animals have a different type of digestive system that enables them to tolerate fungi and other plants which are toxic or deadly to humans.

Even with species of mushrooms generally known to be edible, whether wild or the commercially grown ones sold in the markets, there is always the possibility of personal sensitivity, just as with any other kind of food. We strongly recommend that anyone trying a new species eat it very sparingly at first, so that if any allergy or illness should ensue, the cause could be pinpointed. There are also some mushrooms that are poisonous if eaten raw, but may become harmless when properly cooked.

The two most serious types of mushroom poisoning are due to *amanitin* and *gyromitrin* toxins, which cause blood-cell destruction and severe liver and kidney damage (or failure), and therefore can often result in death. Amanitin is found in many species of the genera *Amanita* and *Galerina* and has also been recorded in *Conocybe* and *Lepiota*. Gyromitrin is reported in some species of *Gyromitra*, *Helvella*, *Morchella*, *Peziza*, *Sarcosphaera*, and *Verpa*.

The most insidious feature of amanitin poisoning is its delayed onset. Gastrointestinal symptoms usually do not appear until 10 to 20 hours after ingestion, and by that time the liver and kidney damage is far advanced. *Amanita phalloides*—a very tasty mushroom, according to its survivors—was once known to occur mostly in the eastern states and Europe; but it has now become well established on the Pacific Coast, particularly in California. It is responsible for most of our mushroom fatalities if much is consumed and proper treatment is not soon available. Its characters, and those of other highly poisonous mushrooms, will be described in detail in the text, and they should be carefully studied by any novice mushroom hunter. Some species of *Galerina* and *Conocybe* also contain amanitin and can be equally deadly, but they are not as conspicuous and fleshy as most of the Amanitas.

Gyromitrin toxin, technically known as *monomethylhydrazine*, is even present in some of our most prized edibles, the morels (genus *Morchella*), but the volatile poison can be dispersed in the steam if the mushrooms are cooked thoroughly in an open pot. (The cook should take care not to inhale these fumes.) The so-called "false morels" (genera

Helvella and *Gyromitra*) contain several species which are highly poisonous, sometimes even after cooking. *G. esculenta*, despite its inviting name, is particularly harmful, and has caused many deaths; symptoms of abdominal pain, gastrointestinal disturbance, and headache may appear from 1 to 24 hours after ingestion.

Many fungi contain gastrointestinal irritants, and reactions to them may vary. Like the preceding group, some may be edible after cooking but poisonous when raw. Most of these highly toxic species are found in the genera *Entoloma*, *Naematoloma*, and *Hebeloma*.

The alkaloid *muscarine* is present in several species of *Amanita*, *Clitocybe*, and *Inocybe*, and it is suspected in some others, such as *Boletus*, *Hebeloma*, *Mycena*, *Omphalotus*, and *Russula*. Symptoms of muscarine poisoning appear quickly, within 20 to 30 minutes, and include excessive sweating and salivation, contracted pupils, diarrhea, and shock.

Coprine poisoning is a rather strange phenomenon because this substance is found only in *Coprinus atramentarius*. This mushroom is highly regarded as an edible, but if alcohol is consumed at the same time, there is usually an unpleasant sensation of heat, marked vascular dilation, and flushing of the skin. Oddly enough, the vasomotor symptoms often reappear for one or two subsequent days if alcohol is taken again, even if no more of the mushrooms are eaten.

Another type of poisoning, caused by *ibotenic acid*, *muscimol*, and related compounds, is evident within 30 to 60 minutes after ingestion of *Amanita muscaria* and *A. pantherina*—mental confusion, visual distortions, and various degrees of euphoria, etc., soon follow. These mushrooms are sometimes deliberately used for these sensations, but there are reports of convulsions and coma in children who have accidentally eaten them.

The hallucinogenic mushrooms are found mostly in the genera *Psilocybe* and *Panaeolus*. They have been used in religious rites by Mexican and Central American Indians for a long time; and in recent years an increasing interest in using these mushrooms for recreational purposes has been noted. They contain either *psilocybin* or *psilocin*, or both; since 1970 these have been on the list of controlled substances and are illegal to possess.

Although not native to the Pacific Coast states, *Psilocybe cubensis* is easily grown in culture and seems to be widely used in certain circles. Symptoms may start within 20 minutes after ingestion, and can range from euphoric sensations and intensified vividness of colors to panic, loss of reality, and fear of impending death.

While emphasis has been placed on poisonous species here to warn the amateur of potential dangers, there are many fine edible mushrooms. Following is a list of some of the most desirable species:

Morchella spp.	*Lycoperdon* spp.
Dentinum repandum	*Calvatia* spp.
Hericium erinaceus	*Agaricus campestris*
Sparassis radicata	*Hygrophorus chrysodon*
Cantharellus cibarius	*Coprinus comatus*
Cantharellus subalbidus	*Pleurotus ostreatus*
Gomphus clavatus	*Lepista nuda*
Craterellus cornucopioides	*Lactarius deliciosus*
Boletus edulis	*Lactarius sanguifluus*

NOMENCLATURE AND CLASSIFICATION

Fungi, like all plants, are classified by mycologists into systematic categories, although much still remains to be learned about their origin and relationships. Most authorities believe that they have probably evolved from algae, a primitive group of chlorophyll-containing plants.

A code of rules for names of the higher plants was proposed by Linnaeus, a Swedish botanist, in 1753, and it remains the basis for our International Rules of Botanical Nomenclature. This binomial, or two-named, arrangement requires that every plant be given a generic name followed by a specific epithet. For fungi, the starting point for official publication is considered to be 1821, with Elias Fries' *Systema Mycologicum*. When a new genus or species is described, it must be published in Latin in a recognized scientific publication to be considered valid.

The order of classification of plants starts with major divisions called *phyla* and descends through the following steps:

PHYLUM (pl. PHYLA)
  CLASS
    ORDER
      FAMILY
        GENUS (pl. GENERA)
          SPECIES

The true fungi belong to the phylum Eumycophyta, which in turn is divided into four classes, only two of which contain fleshy fungi. These are the class Ascomycetes, whose spores are borne in a sac called an *ascus*, and the class Basidiomycetes, whose spores mature on a clublike structure called a *basidium*. The Ascomycetes are divided into 22 orders and the Basidiomycetes into 14 orders. We have not included descriptions of any orders, but a family description is given when it is represented by species in this guide.

There are certain standard word endings for only two of the above categories: *-ales* for an order and *-aceae* for a family. Genus and species names are always either in Latin or in a Latinate form, if the name was derived from Greek or other languages. The genus is capitalized, but in modern usage the species or varietal names are not. Some older literature will show capitalization when the name of a person or locality was used; the correct form now is *Boletus smithii*, *Gyromitra californica*, etc.

The precise definition of a species (the word is the same for both singular and plural) may be modified by *varieties* if there should be different color forms, presence or lack of a distinctive odor, a highly unusual habitat, etc.

In technical works, the name (often abbreviated) of the person who originally described a particular species is written after its Latin name. If the fungus should later be transferred to another genus, the name of the describer is placed in parentheses, followed by the name of the authority making the change. Often two mycologists will independently describe the same mushroom, and when this happens the earlier publication naturally has precedence. If the later name was also commonly used for some time, or had appeared in several publications, it is listed below the correct name and called a synonym. The abbreviations used here for the describers of fungi are listed at the end of this guide.

There are relatively few common names for North American mushrooms, such as "Meadow Mushroom" (*Agaricus campestris*), "Shaggy Mane" (*Coprinus comatus*), "Giant Puffball" (*Calvatia gigantea*), etc. We have mentioned only those common names which are widely recognized and used, because many localities have their own popular name for a particular species, and that same species may be known by various other names in other areas. Obviously, use of the correct scientific name is the only sure way of avoiding confusion.

HOW TO USE A BOTANICAL KEY

A key consists of a series of two or more well-defined choices regarding the characters of the mushroom to be identified. We have included family and genus keys, and in Chapter 4 we give a spore-color key for the agarics. Whenever possible, we have used *macroscopic* descriptions rather than the more technical *microscopic* features. Page numbers of the description of a family, genus, or species follow the names in the keys.

For example, the key to species of the genus *Helvella* on p. 28 gives two choices for the shape of the pileus. If the specimen you want to identify is "cup-shaped" rather than "lobed or saddle-shaped," you would proceed to 2, where you will find that color determines the species: if grayish-black, it is *H. leucomeleana*; but if tan to brownish, it is *H. acetabulum*. However, if the pileus is "lobed or saddle-shaped," you would refer to 3, which indicates that the appearance of the stipe is of prime consideration: if smooth, it is *H. elastica*; but if the stipe is "ribbed or lacunose," you must continue on to 4 to make the color differentiation between *H. lacunosa* and *H. crispa*.

2. ASCOMYCETES/Sac Fungi

Cup, sponge, and saddle fungi are fleshy members of a very large class called the Ascomycetes, or sac fungi. Their spores are produced in saclike structures called *asci*. Each ascus usually contains 8 spores, but in some species the number may be 4 or only 2. Some members of this class, such as the morels, are excellent edibles, but some of the saddle fungi are poisonous. The majority of the Ascomycetes are microscopic, comprising such groups as the yeasts, powdery mildews, molds, ergots, and many others.

Representatives of some of the more common fleshy Ascomycetes are described.

CUP FUNGI, SADDLE FUNGI, AND MORELS

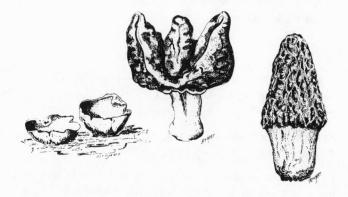

FAMILY HELIOTACEAE

The Heliotaceae contains a number of small cup fungi that grow on fruits, catkins, twigs, and stems of plants as well as on decomposing wood. Some are leathery, others are fleshy, cartilaginous, or gelatinous.

Genus *Chlorociboria*

There is only one common species of *Chlorociboria* in North America.

Chlorociboria aeruginascens (Nyl.) Kanouse Plate 1

CUP up to 0.5 cm in diameter; saucer-shaped to flat; greenish-blue. STIPE very short. SPORES colorless; narrowly ellipsoid; smooth; 5-7 x 2-2.5 μ. EDIBILITY unknown.

C. aeruginascens is widely distributed in the Northwestern states. It grows on dead wood, usually oak, which is often stained green by the mycelium.

FAMILY SARCOSCYPHACEAE

The Sarcoscyphaceae includes a number of cup fungi that are either tough and leathery or gelatinous. These frequently grow on wood and often have a stipe. The fruiting body is called an *apothecium*.

Genus *Sarcoscypha*

Sarcoscypha contains some colorful species which are usually found on rotting wood or fallen branches. The margin of the cup lacks conspicuous hairs.

Sarcoscypha coccinea (Scop. ex Fr.) Lamb.   Plate 2
Plectanea coccinea (Scop.) Fckl.

FRUITING BODY deep cup-shaped; up to 2.5 cm in diameter; inner surface brilliant scarlet; externally whitish and somewhat floccose. STIPE, when present, stout and up to 2 cm long. SPORES colorless; ellipsoid; smooth; 26-40 x 10-12 μ. EDIBILITY unknown.

S. coccinea is one of the most brilliant and conspicuous of cup fungi, common over much of the northern hemisphere. It is widely distributed in western North America. It grows on decaying branches in damp places on the forest floor.

Genus *Plectania*

The fruiting body in *Plectania* is cup-shaped and blackish-brown, with a short stipe.

Plectania nannfeldtii Korf

FRUITING BODY shallowly cup-shaped; up to 3 cm in diameter; blackish-brown with hairs on the outer surface. STIPE up to 4 cm long; rather slender; similar in color to outer surface of cup; hairy; mycelium at base black. SPORES colorless; ellipsoid; smooth; 30-35 x 15 μ. EDIBILITY unknown.

P. nannfeldtii is found at higher elevations in our western coniferous forests, where it grows on rotting wood, often in or near snow.

Genus *Discina*

Members of *Discina* bear a vague resemblance to some species of *Helvella* (see below). The fruiting body is discoid and sometimes irregularly wrinkled. There may be a short stipe. Spores are ornamented at each end, usually with an *apiculus*, a short triangular projection.

Discina perlata Fr.
Acetabula ancilis (Pers. ex Fr.) Lamb.

CUP up to 10 cm in diameter and discoid; surface uneven or sometimes veined; light to dark brown above; whitish beneath. STIPE, if present, 1-3 cm long and equally thick; sometimes lacunose. SPORES colorless; ellipsoid with pointed ends; minute warts on surface; 30-35 x 12-13 μ. EDIBLE.

D. perlata is a common species in spring in the higher mountains of the West. It appears on the ground, often close to rotting wood, shortly after the snow melts and is part of the snowbank mycoflora along with *Lyophyllum montanum* and *Gyromitra gigas*.

FAMILY PEZIZACEAE
(Common Cup Fungi, Saddle Fungi, and Morels)

The Pezizaceae contains the great majority of the cup-shaped fungi, as well as the morels and saddle fungi. Its members are usually fleshy or cartilaginous. In the Sierra Nevada and Rocky Mountains, cup fungi and morels typically fruit in the spring and early summer months. In Alaska they are most abundant in July. Along the Pacific Coast, cup and saddle fungi occur in late fall and winter.

Genus *Sarcosphaera*

Sarcosphaera is a rather large, partly *hypogeous* (fruiting underground) genus of fungi that emerges from the ground and then splits open to expose the inner spore-bearing surface.

Sarcosphaera crassa (Santi ex Steudl) Pouzar
Sarcosphaera coronaria (Jacq. ex Cke.) Boud.

CUP up to 12 cm in diameter; irregularly spherical at first, but soon rupturing with a series of raylike projections; outer surface white; inner spore-bearing surface lilac-gray. SPORES colorless; ellipsoid with ends truncate; smooth; 15-18 x 8-9 μ. POISONOUS.

S. crassa is a common and rather striking cup fungus that is often associated with *Morchella angusticeps* and *Calvatia sculpta* in the higher mountains of the West after the snow melts in late spring and early summer. Sometimes, when just pushing out of the ground, it may be mistaken for a puffball—but unlike the latter, it is hollow within and soon bursts open to expose the delicately lilac-tinted hymenial surface.

Genus *Peziza*

Species of *Peziza* are distributed worldwide and occur commonly on the ground or rarely on wood, usually in damp places. The fruiting body is cup-shaped and, in most species, medium to relatively large and brownish.

Peziza badia Pers. ex Merat

FRUITING BODY cup-shaped; up to 10 cm in diameter; solitary to gregarious; tan to dark brown at maturity; flesh relatively thin, whitish near base. SPORES colorless, ellipsoid; smooth; 17-23 x 8-10 μ. EDIBLE.

P. badia is a common species often seen on the ground in woods or occasionally in the open.

Another fairly common and somewhat similar edible species is **P. repandum** Pers., widespread in coniferous areas and woodlands. This species, however, typically grows on rotten wood and as it matures folds back, thus losing the cup shape possessed when young.

Peziza vesiculosa Fr.
Pustularia vesiculosa (Bull. ex Fr.) Fckl.

FRUITING BODY usually gregarious or densely *caespitose* (clustered), forming a mass of deep cups of irregular shapes; clumps may attain a diameter of up to 15 cm; brown on inner surface; much paler externally, becoming whitish with age; external surface marked by small wartlike processes. A very short stipelike base, often, but not always, present.

SPORES colorless; ellipsoid; smooth; 20-24 x 11-14 μ. EDI-BILITY unknown.

P. vesiculosa is usually found in rich leafmold or areas where manure is present.

A related species, **P. violacea** Pers. ex Fr., is closed at first and opens gradually. This subglobose, deep violet fungus often grows on burnt ground, such as old campfire beds, and on the margin of snowbanks. Its edibility is also unknown.

Genus *Scutellinia*

The fruiting body in members of *Scutellinia* is small and saucer-shaped, with brown hairs around the margin.

Scutellinia scutellata (Fr.) Lamb.
Patella scutellata (Bull. ex Fr.) Fckl.

CUP up to 2 cm in diameter; globose at first, but expanding to a disc; bright red with conspicuous brown hairs around margin of cup. SPORES colorless; ellipsoid; filled with oil drops; minutely sculptured; 20-24 x 12-15 μ. EDIBILITY unknown.

S. scutellata is widespread on rotting wood in forested parts of North America. In parts of Alaska it occurs on humus in the tundra. The fruiting bodies are usually quite gregarious, but can easily be overlooked because of their small size.

Genus *Helvella*

The pileus in members of *Helvella* may be saddle-shaped, lobed, or cup-shaped. A stipe—which may be smooth or ribbed, long or short, hollow or solid—is present. The spores, which are essentially colorless, contain a conspicuous oil drop and are therefore said to be *guttulate*.

Spores of the closely related genus *Gyromitra* (see below)

contain 2 oil drops, but the pileus is generally more con-voluted and brainlike than in species of *Helvella*. Species formerly placed in the genus *Paxina*, such as *H. acetabulum* and *H. leucomelaena*, may first be identified as cup fungi because of their shape. However, in macroscopic charac-teristics they represent forms of *Helvella* in which the spore-bearing structure has not folded backward. If their cups were rolled downward, *H. acetabulum* would somewhat resemble *Gyromitra californica*, and *H. leucomelaena* would look much like a small *H. lacunosa* (see below for all).

No members of *Helvella* should be eaten raw, and it is advisable to parboil even the edible species and pour off the water before final cooking by any method. Never use old fruiting bodies.

KEY TO SPECIES OF *HELVELLA*

1a. Pileus cup-shaped 2
 b. Pileus lobed or saddle-shaped 3
  2a. Pileus grayish-black ..... *H. leucomelaena*, p. 32
   b. Pileus tan to brownish .... *H. acetabulum*, p. 29
3a. Stipe not ribbed or lacunose ..... *H. elastica*, p. 30
 b. Stipe ribbed or lacunose 4
  4a. Pileus blackish to gray .... *H. lacunosa*, p. 30
   b. Pileus and stipe white to
     cream-color .... *H. crispa*, p. 32

Helvella acetabulum (L. ex St.-Amens) Quél.
Paxina acetabulum (L. ex Fr.) Kuntze

PILEUS up to 8 cm in diameter; deeply cup-shaped; brown-ish-tan, somewhat darker on inner surface; outer surface strongly ribbed. STIPE up to 5 cm long; lacunose and strongly ribbed, the ribs continuous with undersurface of pileus; usually tapering toward base; whitish. SPORES color-less; ellipsoid; smooth; with one fairly large central oil drop; 18-22 x 12-14 μ. POISONOUS when raw.

H. acetabulum, commonly termed *Paxina acetabulum* in older literature, is widespread in North America as well as Europe. It grows in both coniferous and deciduous woodland regions, principally in spring and early summer. Its light brown color and the lacunose stipe, with ribs extending up to the margin of the cup in places, make recognition an easy matter.

Helvella elastica Bull. ex Fr.   Plate 3

PILEUS up to 3 cm broad; irregularly saddle-shaped or with 2 or 3 lobes; margin free from stipe, although it may be rolled around it; grayish to olive-brown above, pallid below. STIPE up to 10 cm long; slender; hollow; smooth to slightly wrinkled; white to yellowish. SPORES colorless; ellipsoid; smooth; with one fairly large central oil drop; 10-12 x 18-22 μ. EDIBILITY unknown.

H. elastica is widespread in the western states and provinces. It grows in summer and autumn in moist woods, both coniferous and deciduous; its small size, however, makes it inconspicuous and hence very often overlooked.

Helvella lacunosa Afz. ex Fr.

PILEUS up to 5 cm in diameter, occasionally larger; irregularly lobed and somewhat saddle-shaped; margin reflexed, attached to stipe in places; surface dark gray to black; flesh thin, brittle, whitish. STIPE up to 10 cm long, occasionally longer; tapering upward; deeply ribbed, lacunose; ribs extending onto underside of pileus; white to gray. SPORES colorless; smooth; ellipsoid; with one large central oil drop; 17-20 x 11-13 μ. EDIBLE, but see below.

H. lacunosa (Fig. 7), with its blackish, lobed pileus and white ribbed stipe, is widespread and easily recognizable. It is common in the summer in the Rocky Mountain states and provinces, and in fall and winter along the Pacific Coast. In the Cascades and Sierra Nevada it may be found in late

FIG. 7. *Helvella lacunosa.*

spring and early summer. It grows in coniferous forests, under live oaks, in deciduous woodlands, and even in grassy areas close to trees. Very large fruiting bodies are occasionally found; these may attain a height of 20 cm, but most specimens are much smaller.

Many persons consider this species edible and choice but, as with all members of this genus, one should be cautious. Only young specimens should be selected for eating, and they should be cooked the day they are collected. The stipe is rather tough and rubbery and is best discarded.

H. crispa Scop. ex Fr., not common in the West, somewhat resembles *H. lacunosa* in shape of pileus and in its
relatively long, ribbed, and lacunose stipe. However, the
pileus is buff to umber-brown and the stipe is pubescent. Its
edibility is unknown.

Helvella leucomelaena (Pers.) Nannf.
Paxina leucomelas (Pers.) Kuntze

PILEUS up to 3 cm in diameter; deeply cup-shaped; outer
portion blackish-brown above, becoming white toward stipe;
margin whitish, occasionally serrate; inner portion of cup
blackish with bloom from spores when mature. STIPE up to
4 cm long; 0.5 cm in diameter; ribbed, deeply lacunose, with
ribs extending onto base of cup; white. SPORES colorless;
ellipsoid; smooth; with one very large central oil drop;
18-22 x 10-13 μ. EDIBILITY unknown.

H. leucomelaena is uncommon but has been recorded in
California, Alaska, and the Rocky Mountains. It is rather
inconspicuous and may be found growing in soil under
conifers; the white stipe is usually embedded in the ground.

Genus *Gyromitra*

The fruiting body in *Gyromitra* is stipitate and pileate,
with the brown or brownish-red pileus usually wrinkled,
folded, and occasionally saddle-shaped. The stipe usually
has internal folds. Spores are colorless, narrowly ellipsoid,
smooth or finely warted, and usually have 2 oil drops.

The same precautions should be taken with members of
Gyromitra as is indicated in *Helvella* (see above). Even the
edible *G. gigas* should first be parboiled in an open pot.

Gyromitra californica (Phill.) Raitv.   Plate 4
Helvella californica Phill.

PILEUS up to 10 cm in diameter, usually less; reflexed;
lobed; margin free (not attached to stipe), extending down-

ward to cover much of stipe; light brown with olive tinge above; underside white or creamy-white. STIPE up to 8 cm long; 3-5 cm in diameter; hollow; deeply ribbed; lacunose with ribs extending upward, continuing onto underside of pileus; surface dry, creamy-white above, becoming pink toward base. SPORES colorless; ellipsoid; smooth; with small oil drop on each end; 16.5-17.5 x 8-9.5 μ. POISONOUS.

G. *californica*, one of the most beautiful members of *Gyromitra*, occurs in the higher mountains of the West in late spring and summer. Favored habitats are along roadsides in coniferous forests, when the soil is moist, or in grassy patches along small streams. The pileus often obscures the delicate pink, ribbed stipe until one digs up the fruiting body.

Gyromitra infula (Schaeff. ex Fr.) Quél.   Plate 5
Helvella infula Schaeff. ex Fr.

PILEUS up to 10 cm in breadth; saddle-shaped, but lobes often very irregular; margin not attached to stipe; surface smooth to irregularly convoluted; reddish-brown. STIPE up to 7 cm long; 1-2 cm in diameter; hollow, usually slightly enlarged at the base; surface sometimes has a few folds; whitish to pale brown with vinaceous tinge. SPORES colorless; sub-ellipsoid; smooth; with 2 large oil droplets. 18-20 x 7.5-8.5 μ. POISONOUS; see below.

G. *infula* is found most frequently in summer and fall on rich soil or rotten wood, usually in coniferous forests. It is widely distributed along the Pacific Coast and in the Rocky Mountains. It is easily identified by the dark brown, saddle-shaped pileus, without conspicuous convolutions, and the hollow, unribbed stipe. Thin sections of the pileus turn pinkish-red when treated with 2-percent KOH. Do not eat this species! It belongs to a group containing toxins that affect the blood cells as well as the central nervous system.

Gyromitra gigas (Krombh.) Quél.
Helvella gigas Krombh.

PILEUS up to 20 cm in diameter; surface folded; yellowish-brown. STIPE relatively short but massive toward base; deeply folded; white. SPORES colorless; ellipsoid; mostly smooth but with occasional ornamentation at each end; 24-36 x 10-16 μ. EDIBLE and choice; see below.

   G. gigas is a common snowbank species in the higher mountains of the West. It appears in late spring or early summer as the snow melts; in parts of the Rocky Mountains it is known as the Brain Mushroom or Snow Mushroom. It is edible and choice and highly sought after, especially in Idaho. Since a single fruiting body may weigh up to one pound or even more, a clump of *G. gigas* provides a lot of food; it may be fried in butter or margarine like morels and frozen for future consumption.

   A closely related species, **G. fastigiata** (Krombh.) Rehm, occurs at lower elevations. The pileus is darker brown and more convoluted. It is reported to be poisonous.

Gyromitra esculenta Fr.   Plate 6

PILEUS up to 8 cm in diameter; irregularly lobed; surface very wrinkled, with brainlike convolutions; light to dark reddish-brown. STIPE up to 5 cm long; moderately thick; surface nearly smooth; slightly wrinkled; whitish to pale tan; hollow; sometimes compressed. SPORES colorless; ellipsoid; smooth; 17-28 x 7-16 μ. POISONOUS.

   G. esculenta appears in late spring and summer in the coniferous forests of the West from Alaska southward. The numerous fine convolutions on the pileus distinguish it from *G. infula* and *G. gigas* (see description of both above). We strongly advise against eating *G. esculenta*, although the poison is said to be removed by parboiling.

Genus *Verpa*

The pileus in *Verpa* is bell-shaped, either smooth or wrinkled; it is easily separated from the stipe, and the flesh is very brittle. The stipe is cylindrical, smooth, hollow, and of nearly equal diameter throughout its length.

Two species of *Verpa* are commonly found in the West; both may be confused with the genera *Morchella* and *Helvella*. However, *Verpa* lacks the pitted pileus with sterile ridges of *Morchella* (see below), and the saddle- or cup-shaped pileus of *Helvella* (see above).

Verpa conica (Müll.) Swartz ex Pers.
Verpa digitaliformis Pers.

PILEUS up to 2 cm in diameter, and of similar depth; campanulate to conic; surface smooth; brown on outer

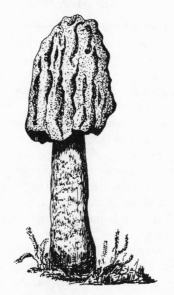

FIG. 8. *Verpa conica.*

surface, white beneath. STIPE up to 6 cm long; cylindrical; hollow; whitish; minutely scaly. SPORES colorless to slightly yellowish; ellipsoid; smooth; 22-26 x 12-16 μ. EDIBILITY unknown.

V. conica (Fig. 8) fruits in the spring and early summer in western North America and is often found in moist grassy situations along streams or at the margins of mountain meadows.

V. bohemica (Krombh.) Schroet. occurs in the same habitats; it is distinguished from *V. conica* by longitudinal ribs on the pileus and by its very large spores, which measure 60-80 μ in length. Its edibility is questionable. Although it is safely eaten in very moderate amounts by many people, eating it on successive days or consuming a large amount at once usually causes lack of muscular coordination.

Genus *Morchella*

Members of *Morchella* are the true morels. The pileus is roughly conic, its surface deeply pitted between cross-veined vertical sterile ridges; it ranges from yellowish or pale gray to dark brown with black ridges. The stipe is light in color, minutely roughened, sometimes slightly wrinkled, and hollow. Spores are ellipsoid and yellowish in mass.

Morels could be confused with species of the genus *Verpa*, especially *V. bohemica* (see above). In *Verpa*, the pileus is attached only to the top of the stipe and hangs down in a skirtlike manner, whereas in all except one species of *Morchella* the pileus is attached directly to the stipe and lacks a free margin; in **M. semilibra** D.C. ex Fr., the attachment of the pileus to the stipe begins about midway from the tip to the lower margin.

Morels are regarded as among the most delicious of fungi. One who finds a locality where they occur can return year after year at approximately the same time and reap a harvest.

Morels do not grow in dense coniferous forests, preferring semi-open situations where some sunlight penetrates. Burnt-over areas produce large numbers of these fungi a year or two after the fire—one famous example was after the Sun-dance fire in northern Idaho a few years ago, when huge numbers of fruiting bodies were collected. In California the fruiting season extends from mid-May to mid-June, depend-ing on the melting of the winter snowpack; in Idaho morels are abundant on the lower mountain slopes dominated by Yellow Pine in mid-May. As the season advances, they appear higher up in fir and spruce forests and may even fruit until July.

Morchella angusticeps Pk.   Plate 7

PILEUS up to 8 cm long; variable in shape, but most often conic; light gray when very young, but ribs gradually be-come blackish on edges and pits darken to brown; attached to stipe throughout its length. STIPE up to 10 cm long; usually enlarged toward base; occasionally wrinkled; surface minutely roughened; hollow; whitish to pale tan. SPORES colorless to pale yellow in mass; ellipsoid; smooth; 24-28 x 12-14 μ. EDIBLE and choice.

M. angusticeps is the most common morel in the West. It occurs widely in coniferous forests at middle elevations; in the Sierra Nevada and Cascades it is commonly associated with Yellow Pine, Jeffrey Pine, and White Fir, especially where there is some undergrowth of *Ceanothus velutinus* or *C. cordulatus*. It ranges north to Alaska and Yukon Terri-tory where it may even grow on the tundra.

The color in *M. angusticeps* is variable; in a cluster growing within a few feet of each other, the ridges on the pileus of large specimens may range from gray to black.

In coastal California, it is not uncommon to find **M. conica** Fr. growing in sandy soil amid low vegetation in the open. This species, now sometimes regarded as a variety of **M. deliciosa** Fr., appears in midwinter and differs from *M.*

angusticeps in its very elongate pileus, ending in an acute apex, and its yellow color; the ribs are generally lighter than the pits. All morels are edible.

Genus *Caloscypha*

The fruiting body in *Caloscypha* is cup-shaped and either *sessile* (lacking a stipe) or stipitate, fleshy or cartilaginous; spores are globose and smooth.

Caloscypha fulgens (Fr.) Fckl.   Plate 8

CUP up to 4 cm in diameter; fairly deep; regular or occasionally unequal sides; inner surface orange-yellow; external surface pale yellow, becoming green around the rim; only a faint indication of a stipe. SPORES colorless; globose; smooth; 6-8 μ in diameter. EDIBILITY unknown.

C. fulgens is abundant in spring in the western mountains. In the Sierra Nevada, it sometimes almost forms mats on the needle-covered ground beneath Western Yellow Pines at the time these trees are shedding their pollen. As a result, it is often difficult to photograph specimens that are not heavily dusted with yellow powder. The fruiting bodies may be scattered or crowded together in a rosette formation.

Genus *Aleuria*

The fruiting body in *Aleuria* is cup-shaped to flat or irregularly folded; the external surface is bright orange to red, smooth or with a whitish *tomentum* (matted wooly hairs), and the flesh is thin.

Aleuria aurantia (Fr.) Fckl.

CUP up to 10 cm in diameter; usually irregularly saucer-shaped, but when caespitose they may be compressed from pressure; upper or hymenial surface bright orange; external surface whitish; flesh thin and brittle. SPORES colorless; ellipsoid; at first smooth, then becoming sculptured in the form of reticulations; 18-22 x 9-10 μ. EDIBLE.

A. aurantia, often called the Golden Cup Fungus, is probably the most common and widespread species of cup fungus in North America. It shows a preference for old dirt roads or roadbanks where the soil is packed and often of poor quality, appearing during the summer months in the Rocky Mountains and from fall to spring along the Pacific Coast and in the Sierra foothills. It is among the easiest of fungi to identify.

Genus *Otidea*

The fruiting body in *Otidea* is single to caespitose, usually asymmetrical with a split on one side, and so elongated as to appear ear-shaped.

Otidea leporina (Fr.) Fckl.
Scodellina leporina (Batsch ex Fr.) S. F. Gray

CUP up to 5 cm high and 2-3 cm wide, with one side slit to base; yellowish-brown to wood-brown. SPORES colorless; narrowly ellipsoid; smooth; with 2 small oil drops; 12-14 x 7-8 μ. EDIBILITY unknown.

O. leporina literally means rabbit-ear, which very well describes the appearance of this cup fungus. It is fairly common on the ground in forested areas in fall and winter along the Pacific Coast, and in spring in the higher mountains of the West. There are several other species of this genus in western North America but all superficially resemble one another.

EARTH-TONGUES, TRUFFLES, AND OTHER ASCOMYCETES

FAMILY GEOGLOSSACEAE

The fruiting body in the Geoglossaceae—which is Latin for earth-tongues—may be club-shaped, spathulate, or pileate. Some species superficially resemble certain members of the families Clavariaceae and Cantharellaceae among the Basidiomycetes (see next chapter), while others may resemble a *Helvella* (see above, under Cup Fungi, etc.). The Geoglossaceae, however, has many distinctive structural characters. Spores vary in shape from ellipsoid to filiform or whip-shaped, and may or may not be septate; in some species the spores may contain one or more oil drops.

Ten genera are recognized, only a few of which will be discussed here. None of the Geoglossaceae is particularly common in the West, and they are of no importance to those whose interest in fungi pertains only to edible species.

Genus *Trichoglossum*

The fruiting body in *Trichoglossum* is black and more or less lanceolate, with a basal and non-spore-bearing stipe. Minute *setae* (bristles) are scattered among the asci.

Another genus, **Microglossum**, resembles *Trichoglossum* in size and shape, but its members are usually brighter in

color, often green, yellow, or reddish-brown. A bright green
species, **M. viride** Fr., occurs in California on the ground in
moist woods. Its edibility is unknown.

Trichoglossum hirsutum (Fr.) Boud. Plate 9

FRUITING BODY up to 7.5 cm in height; upper, spore-bearing
third ellipsoid or lanceolate; compressed, sometimes longitudinally grooved; black; surface contains numerous minute
brown spines. Lower two-thirds, constituting STIPE, about
0.3 cm in diameter; round in cross-section. SPORES brown;
very elongate, tapering at both ends; 15 septa when mature;
100-150 x 6-7 μ. EDIBILITY unknown.

T. hirsutum, a shining black fungus with an oblong,
stalked fruiting body, is striking in appearance but sometimes difficult to see amid the humus of the forest floor
where it grows. It is widespread in the Northwest and
coastal states in autumn and winter in both coniferous and
nonconiferous forests; though not common, it is usually
quite gregarious when it occurs.

Genus *Spathularia*

Spathularia somewhat resembles *Trichoglossum* and *Microglossum* (see both above) in shape, but the fruiting
bodies have a compressed, lanceolate, or oblong pileus that
is sharply delimited from the stipe and decurrent on opposite sides.

Spathularia flavida Fr.

FRUITING BODY up to 10 cm in height; spores produced on
upper third, which is oblong to lanceolate, occasionally
irregular in shape, compressed, and distinctly marked off
from stipe, on which it is decurrent on opposite sides; pallid
to yellow or brownish with age. Lower two-thirds, constituting STIPE, round to compressed in cross-section; often

expanding toward base. SPORES colorless; very elongate, almost filiform; smooth; with a number of septa; 35-40 x 2 μ. EDIBLE.

S. flavida is another small Ascomycete of the forest floor that one is apt to see while down on the ground looking for something else.

Genus *Mitrula*

Mitrula is similar to *Spathularia* and *Microglossum* (see descriptions of both above) in shape and yellowish color, but differs from both these genera in habitat and in not having multiseptate spores.

Mitrula paludosa Fr.

FRUITING BODY up to 5 cm in height, rarely higher; upper fifth bearing spores, marked off from stipe; spore-bearing part elliptical to fan-shaped, eggyolk-yellow. Lower four-fifths slender, stipelike; white; viscid. SPORES colorless; cylindrical; smooth; 10-18 x 2.5-3 μ. EDIBILITY unknown.

M. paludosa is a species of the spring or early summer in the West. This small but colorful fungus most often grows on decaying vegetation, fallen pine needles, or twigs on the edge of vernal pools, springs, or shallow ponds left from melting snow. The bright yellow spore-bearing head on a slender white stalk are good field characters for identification.

Genus *Leotia*

The fruiting body in *Leotia* has a pileus and stipe some-what resembling certain species of *Helvella* (see above, under Cup Fungi, etc.). The flesh in *Leotia*, however, is gelatinous, and the spores have 3 to 5 septa.

Leotia lubrica Fr.

PILEUS up to 2.5 cm in diameter; roughly hemispheric, but irregularly lobed and wrinkled; olive-yellow to bright yellow, sometimes with a greenish tinge; gelatinous; extremely viscid. STIPE up to 7 cm long; round in cross-section, occasionally compressed; tapering upward slightly; similar in color to pileus or slightly more orange-yellow; viscid. SPORES colorless; narrowly ellipsoid to subfusiform; smooth; with 3 to 8 oil drops; 5 septa when mature; 16-23 x 4-6 μ. EDIBILITY unknown.

L. lubrica occurs worldwide, frequently on the ground or on rotting wood in fall and winter in forested parts of western North America.

L. viscosa Fr., which has a dark green pileus and a white to orange stipe, has also been found in the West. Its edibility is unknown.

Genus *Cudonia*

Members of *Cudonia* are somewhat similar in shape to species of the genus *Leotia* (see above), but are fleshy or even leathery rather than gelatinous, and do not have a viscid surface. Only the upper surface bears spores.

Cudonia circinans Fr.

PILEUS up to 2 cm in diameter; usually slightly convoluted or lobed; fleshy; cream-color to brown. STIPE up to 5 cm long; round or slightly compressed; sometimes hollow; pallid to vinaceous-brown; surface usually slightly scurfy. SPORES colorless; acicular (needle-shaped); smooth; 32-40 x 6 μ. EDIBILITY unknown.

C. circinans is a species one is not likely to encounter frequently, but it has been found in Oregon, Washington, and Idaho, and may be quite widespread. At first glance it looks like a small species of *Helvella* (see above, under Cup Fungi, etc.), but an examination of the acicular spores immediately eliminates it from that genus.

FAMILY HYPOCREACEAE

The Hypocreaceae is a small family of parasitic fungi whose mycelia grow in association with those of other fungi. At maturity the fruiting body covers that of its host as a thin soft layer on which the *perithecia* (small flasklike structures containing the asci) are borne. Each ascus contains 8 two-celled spores.

Genus *Hypomyces*

There are six species of *Hypomyces* in North America, and all are parasitic on other fungi.

Hypomyces lactifluorum (Schw.) Tul.   Plate 10

FRUITING BODY covers the pileus, gills, and most of stipe of host; bright orange-red; surface minutely dotted with perithecial openings. SPORES colorless; fusiform; rough; 30-40 x 6-8 μ. EDIBLE, but not recommended; see below.

H. lactifluorum is a common parasite on members of the genera *Lactarius* and *Russula*, especially *R. brevipes* (see below, under Basidiomycetes) in the western states and provinces. It so completely covers the surface of the fruiting body of its host that it usually stunts the growth of the gills and causes them to be only faintly indicated as ridges. Although known to be edible, this parasite could possibly occur on a poisonous mushroom that might not be easily identifiable.

FAMILY TUBERACEAE/TRUFFLES

Truffles belong to an order of Ascomycetes called the Tuberales. They are *hypogeous* (fruiting underground) fungi, hence difficult to find. Although they were famous in Europe and sought as a delicacy for several thousand years, only two species were known from America until 1899, when Dr. H. W. Harkness published a paper describing a number of new genera and species from western North America in the *Proceedings of the California Academy of Sciences*. Since then there has been considerable interest in this group in this country. Although few collectors are likely to find truffles, we felt that the general interest in these subterranean fungi warranted the description of one species which has been found from central California north to British Columbia.

The fruiting body in the family Tuberaceae is usually roundish, and the hymenium is folded into a brainlike structure or series of chambers. The mycelium of most truffles forms a mycorrhizal relationship with the roots of trees, shrubs, or even herbaceous plants.

Genus *Tuber*

The fruiting body in the genus *Tuber* is round to irregular and fleshy or cartilaginous, with a smooth or rough surface. The asci range from round to pear-shaped or ellipsoid and contain 1 to 4 or more round or ellipsoid spores.

Tuber gibbosum Hark.

FRUITING BODY up to 5.5 cm in diameter; subglobose and often lobed; surface with matted or parallel hairlike cells; light buff to brown. GLEBA (spore mass) yellowish when young, becoming reddish-brown at maturity; marbled with white veins; often with open spaces. SPORES yellowish-brown; ellipsoid; ornamented; 35-52 x 17-39 μ. EDIBILITY unknown.

T. gibbosum has been collected along the Pacific Coast in the San Francisco Bay area and northern California, Oregon, and southern British Columbia. Helen Gilkey (1939) described and named a number of other species of *Tuber* from the Pacific Coast. Because of their rarity, we know nothing about their edibility.

3. BASIDIOMYCETES/
Club Fungi
Nongilled Fungi

In the second great class of fleshy fungi, the Basidio-mycetes, the spores are borne on the tips of elongate, often club-shaped structures called *basidia* on the hymenial (spore-producing) surfaces of the fruiting body. The basidia may be on the gills of gill fungi, within the pores of boletes or polypores, on the surface of jelly fungi, or on the "teeth" of teeth fungi. The spores may be borne in a glebal mass and remain encased within the outer surface of the fruiting body until they are mature and powdery, as with puffballs; or they may emerge as a slimy mass whose fetid odor attracts insects to disseminate the spores that stick to their bodies, as in the stinkhorns.

We have divided our discussion of the Basidiomycetes into two chapters. This chapter describes the nongilled fungi and the next chapter describes the gilled fungi or agarics.

JELLY FUNGI

FAMILY TREMELLACEAE

The Tremellaceae contains the common jelly fungi, which may be soft and gelatinous or cartilaginous. Some of them are colorful, being brilliant yellow, orange, red or white, and all are associated with cool, wet weather. Microscopically they are distinguished by their basidia, each of which, instead of being a club-shaped structure bearing 4 spores, is split longitudinally into 4 cells, each bearing a single terminal spore.

Genus *Pseudohydnum*

The single species in *Pseudohydnum* is a white, translucent jelly fungus with numerous spinelike processes on the undersurface.

Pseudohydnum gelatinosum
(Scop. ex Fr.) Karst.   Plate 11

Tremellodendron gelatinosum (Scop. ex Fr.) Fr.
Hydnum gelatinosum Scop. ex Fr.

FRUITING BODY up to 6 cm in diameter; often spathulate; surface slightly rough, white to pale tan; translucent; underside of pileus has toothlike structures. Lower part of fruiting body often stipelike. SPORES white; subglobose; smooth; 5-7 μ. EDIBLE.

P. gelatinosum grows on rotting wood in coniferous forests, and is found only during wet weather; it is common in Pacific coastal forests as well as inland mountains of the West. The fruiting bodies may be solitary, gregarious, or even caespitose, and show considerable variation in size. This delicate jelly fungus might at first be mistaken for a tooth fungus (see next section), but all of the latter are fleshy or tough, and none is translucent.

Genus *Tremella*

The fruiting body in *Tremella* is gelatinous and varies in shape from amorphous to lobed, or even branched. Spores are subglobose to ellipsoid and white to colorless.

Members of *Tremella* may be confused with those of the family Dacrymycetaceae (see below) superficially. Microscopic examination, however, will show the latter to have Y-shaped basidia with 2 spores, unlike the longitudinally split basidia characteristic of the true jelly fungus.

Tremella mesenterica (S.F. Gray) Pers.
Tremella lutescens Fr.

FRUITING BODY up to 10 cm in diameter; surface convoluted or lobed; gelatinous and slimy when wet; orange-yellow. SPORES whitish; ovoid; smooth; 10-14 x 7-9 μ. EDIBILITY unknown.

T. mesenterica is often called Witches' Butter. It grows on logs and in crevices in bark, appearing during rainy weather. Within a few dry days after a rain it shrivels and dries into a hard mass which is capable of reviving with another rain. It is widespread in both coniferous and nonconiferous forests in the western states.

Genus *Phlogiotis*

The fruiting body in *Phlogiotis* is terrestrial and spathulate to funnel-shaped, with rubbery flesh. Spores are white and oblong.

Phlogiotis helvelloides (Fr.) Martin   Plate 12
Gyrocephalus rufus (Jacq.) Brefeld

FRUITING BODY up to 8 cm high; spathulate to funnel-shaped; narrowing to a stipelike base; smooth; gelatinous; orange-pink to reddish. SPORES white; oblong; smooth; 10-12 x 4-6 μ. EDIBLE; see below.

P. helvelloides is a beautiful jelly fungus common on moist leaf litter in coniferous forests in the Northwestern and Rocky Mountain states in the fall. Its flattened, tongue-shaped form and rich pink to reddish-orange color readily identify it. It is sometimes candied before eating.

FAMILY AURICULARIACEAE

There are a number of diverse genera in the Auriculariaceae which may be either saprophytic or parasitic on other plants, including mosses, fungi, and vascular species. The fruiting body is not separated into a stipe and pileus, and the basidia are septate. Only one common species that will frequently be encountered by the collector is described.

Genus *Auricularia*

Species of *Auricularia* are rubbery or gelatinous, with free portion cup-shaped or ear-shaped. They are saprophytic on logs.

Auricularia auricula (Hook.) Underw.   Plate 13
Hirneola auricula-judae (Fr.) Berk.

FRUITING BODY up to 15 cm in diameter; ear- or cup-shaped; often slightly ribbed; gelatinous, thin but rubbery; smooth; brownish. SPORES colorless; allantoid; 12-16 x 4-5 μ. EDIBLE.

A. *auricula* grows on decaying logs in coniferous forests. It is fairly common in the western mountains and coastal areas in the fall, and is also sometimes found in late spring at higher elevations. Because of its shape and consistency, it might be confused with the cup fungi which, of course, are Ascomycetes (see preceding chapter).

FAMILY DACRYMYCETACEAE

Some Dacrymycetaceae resemble jelly fungi, while others are shaped like coral fungi, but all are characterized by having 2 spored, Y-shaped basidia.

Genus *Guepiniopsis*

The fruiting body in *Guepiniopsis* is gelatinous, with the spore-bearing surface concave. Spores are allantoid and septate.

Guepiniopsis alpinus (Tracy and Earle) Bres.

FRUITING BODY up to 1.5 cm in diameter; cone-shaped, with apex attached to wood; large end concave and spore-bearing; gelatinous; bright yellow. SPORES allantoid; smooth; septate; 15-18 x 5-6 μ. EDIBILITY unknown.

G. alpinus is a very common snowbank fungus in the higher mountains of the West. It appears on logs and stumps as the snow melts, often growing in rows, with the apex of each fruiting body coming out of a crevice, and the large, spore-bearing end handing down.

TEETH FUNGI

FAMILY HYDNACEAE

The Hydnaceae contains a group of fungi whose spore-bearing surface is composed of toothlike spines directed downward. The fruiting body may be soft and fleshy, leathery, or almost woody. Most species are stipitate, but some have their "teeth" hanging from a fleshy mass attached to wood. The spores range from round to ellipsoid and may be white or brown, smooth or ornamented.

Some taxonomists have divided the Hydnaceae into several separate families but in this guide the members are treated as a single group. The leathery or woody species are inedible, but members of the genera *Dentinum* and *Hericium* are excellent edibles.

KEY TO SOME COMMON GENERA OF HYDNACEAE

1a. Spores white 2
 b. Spores brown.................................. 3
  2a. Fruiting body large, fleshy,
     terrestrial .... *Dentinum*, p. 52
   b. Fruiting body small, tough,
     attached to cone .... *Auriscalpium*, p. 53
3a. Fruiting body growing on wood .... *Hericium*, p. 56
 b. Fruiting body terrestrial 4
  4a. Fruiting body fleshy .... *Hydnum*, p. 54
   b. Fruiting body tough, fibrous.. *Hydnellum*, p. 54

Genus *Dentinum*

The fruiting body in *Dentinum* is large and fleshy, with both pileus and stipe; it is terrestrial.

Dentinum repandum (Fr.) S.F. Gray Plate 14
Hydnum repandum Fr.

PILEUS up to 15 cm in diameter; broadly convex; surface dry; smooth; pale orange-buff. FLESH soft; brittle; cream-color. TEETH from 0.4-0.8 cm long; cream-color. STIPE up to 8 cm long; solid; smooth; whitish. SPORES white; globose; smooth; 6.5-8.5 μ. EDIBLE and choice.

D. repandum is a common species in late summer and fall in the Rocky Mountains, and in the fall and winter along the Pacific Coast. It is most commonly found in coniferous forests, and seems to show no preference—we have found it abundant under Coast Redwood, Douglas Fir, Beach Pine, Bishop Pine, Western Yellow Pine, Lodgepole Pine, and Sitka Spruce, to indicate some of the wide-ranging situations in which it occurs. Because of its abundance, it is excellent in sauces and stews.

Along the Pacific Coast from California northward there is a smaller species, **D. umbilicatum** (Pk.) Pouzar, which has a central depression in the pileus and is smaller and darker. It is equally edible.

Genus *Auriscalpium*

The fruiting body in *Auriscalpium* is small, brown, and stipitate; the stipe is very slender and eccentric.

Auriscalpium vulgare S.F. Gray
Hydnum auriscalpium Fr.

PILEUS up to 2 cm in diameter; surface covered with fine brown fibrils; plane. FLESH thin; firm; tough; white. TEETH gray or violet-gray to brownish with age. STIPE up to 8 cm long; slender; attached to one side of pileus; hairy; dark brown. SPORES white; subglobose; minutely ornamented; 4-5 x 5-6 μ. EDIBILITY unknown.

A. vulgare grows on decaying cones, principally those of Douglas Fir, on the Pacific Coast. It has been reported to occur on pine cones in other areas.

Genus *Hydnum*

The fruiting body in *Hydnum* is fleshy, brittle, and stipitate, with brown, heavily ornamented spores; it is terrestrial.

Hydnum imbricatum Fr.
Sarcodon imbricatus (L. ex Fr.) Karst.

PILEUS up to 25 cm in diameter; somewhat irregular in outline, but broadly convex, with a depressed center; surface covered with coarse, often overlapping brown scales. FLESH thick; rather tough; white; only slightly bitter; does not blacken in KOH. TEETH up to 1.5 cm long; pale gray at first, becoming brownish with age. STIPE up to 9 cm long; relatively thick; smooth; solid; similar in color to teeth. SPORES brown; subglobose; tuberculate; 6-8 x 5-7 μ. EDIBILITY questionable; see below.

H. imbricatum is fairly common in coniferous and mixed forests of western North America, where it ranges northward to Alaska and the Yukon Territory. Its large size and the coarse brown surface of the pileus make recognition easy. There are varying reports of its edibility, and there may be different strains, because some have found it so bitter that it was inedible.

Genus *Hydnellum*

The fruiting body in *Hydnellum* is tough or woody, variable in size; it is terrestrial. Spores are brown, mostly irregular in shape, and tuberculate.

Hydnellum suaveolens
(Scop. ex Fr.) Karst.   Plate 15
Hydnum suaveolens Scop.

PILEUS up to 15 cm in diameter; plane when mature; surface uneven; margin thick; white and somewhat woolly, later becoming grayish-brown. FLESH tough, with a distinctly sweet aromatic odor. TEETH up to 0.8 cm; white at first, but soon becoming dark brown basally and gray at tips. STIPE up to 4 cm long; relatively thick; white at top, shading to bluish-black at base; mycelium blue. SPORES brown; very irregular in shape; tuberculate; 3-3.5 x 3.5-5 μ. EDIBILITY unknown.

H. suaveolens fruits in coniferous forests of the Pacific Coast in fall and early winter, and in the Rocky Mountains in late summer and fall. The soft white pileus of young specimens, combined with the bluish stipe and strong sweet odor, make recognition easy. Like other members of this genus, *H. suaveolens* may envelop itself around other plants. In Plate 15 one can see California Huckleberry (*Vaccinium ovatum*) growing through the pileus of a fruiting body; this picture was taken along the central Oregon coast in October.

Hydnellum aurantiacum (Batsch ex Fr.) Karst.
Hydnum aurantiacum Alb. and Schw.

PILEUS up to 15 cm in diameter; surface plane to depressed and lumpy; whitish when young, but soon bright orange and brown; velvety. FLESH tough; brown to gray; lacking a distinctive odor or taste. TEETH short; white, becoming brown at tips with age. STIPE up to 4 cm long; thick; surface velvety; orange to dark brown. SPORES brown; subglobose; tuberculate; 5.5-7.5 x 5-6 μ. EDIBILITY unknown.

H. aurantiacum is very common in coniferous forests from central California northward along the Pacific Coast in fall and winter, and it is also found in the Rocky Mountains in late summer and early fall. The velvety orange

surface of the pileus often misleads one into thinking it to be
a polypore (see that section, below) until the teeth on the
hymenial surface are seen.

Hydnellum peckii Banker   Plate 16

PILEUS up to 14 cm in diameter; convex or plane; surface at
first finely tomentose; white, exuding red droplets, when
young, becoming brown or gray with age. FLESH soft but
rather tough; brownish. TEETH short; grayish-white, becoming reddish-brown with age. STIPE up to 8 cm long; surface
felty; whitish-gray to reddish-brown; base terminating in a
white mycelioid root. SPORES brown; subglobose; tuberculate; 4.5-5.5 x 3.5-4.5 μ. EDIBILITY unknown.

H. peckii is a striking-looking fungus when young, with
blood-red droplets of juice exuding onto the soft white
surface of the pileus. We have found it in fall under Beach
Pine (*Pinus contorta*) along the Oregon coast; needles of the
pine often become encased in the growing fruiting body.

Genus *Hericium*

The fruiting body in *Hericium* consists of a fleshy mass
attached to dead or living wood. Spines are pendant,
unbranched or branched; spores are white, subglobose or
minutely ornamented.

Hericium erinaceus (Bull. ex Fr.) Pers.   Plate 17
Hydnum caputmedusae Bull. ex Fr.
Hydnum erinaceus Bull. ex Fr.

FRUITING BODY up to 15 or more cm in diameter; often
grows in a spherical mass, but occasionally elongate; covered with pendulant, unbranched spines up to 5 cm long;
pure white, becoming yellowish with age; droplets of clear
liquid often present on tips of spines. SPORES white; round;
smooth; 4.5-6 μ. EDIBLE and choice.

H. erinaceus is commonly called the Hedgehog Fungus because of its long spines (the generic name of the hedgehog of the Old World is *Erinaceus*). It is found along the Pacific Coast in Douglas Fir and Tanbark Oak forests, occurring on both living trees and fallen logs in fall and winter. A single mass we collected weighed 12 pounds. Like other members of *Hericium*, it is edible and choice when young and white.

Hericium coralloides (Scop. ex Fr.) S.F. Gray
Hydnum coralloides Scop. ex Fr.

FRUITING BODY up to 25 cm in diameter; composed of branches with numerous spines, up to 1 cm in length, hanging downward; pure white, becoming yellow on tips of spines with age. SPORES white; round; smooth; 5-6 μ. EDIBLE and choice.

H. coralloides is a very beautiful fungus which grows on both coniferous and hardwood logs in fall and winter along the Pacific Coast. It prefers moist, shady situations and may appear year after year on the same stump or log. It is excellent for eating, but unfortunately is seldom found in quantity.

A closely related edible species, **H. ramosum** (Bull. ex Mérat) Letellier, differs in having shorter, essentially non-pendant spines all along the branches.

CORAL FUNGI

FAMILY CLAVARIACEAE

The Clavariaceae, or coral fungi, comprises a large and diverse group of the Basidiomycetes whose fruiting bodies vary from a single club-shaped stalk to a multi-branched structure somewhat resembling certain kinds of coral. Since these fungi lack gills, tubes, or teeth, the spores are borne on the surface of the clubs or branches and are mostly *hyaline* (colorless), white, or creamy-yellow, but in a few kinds they may be brown or reddish; they vary in shape from nearly round to elliptical, and may be smooth or ornamented. The colors of the fruiting body show great diversity: white, cream, yellow, orange, red, purple, pink, and even green coral fungi can be found in the West.

Along the Pacific Coast, coral fungi appear in autumn and early winter, but in the inland mountains there are both spring and autumn species. A number of kinds are considered edible and good—but there are a few, such as *Ramaria formosa* and *R. gelatinosa*, that may cause illness in some people.

KEY TO SOME COMMON GENERA
OF CLAVARIACEAE

1a. Fruiting body club-shaped,
   unbranched . . . . *Clavariadelphus*, p. 61
  b. Fruiting body not club-shaped, usually branched . . 2
  2a. Branches flattened, arising from a stout,
      rooted stalk . . . . *Sparassis*, p. 69
    b. Branches not flattened 3
3a. Spores yellow to rusty-cinnamon . . . *Ramaria*, p. 66
  b. Spores white 4
  4a. Basidia 2-spored . . . . *Clavulina*, p. 65
    b. Basidia 4-spored 5
5a. Lignicolous and pyxidately
   branched . . . . *Clavicorona*, p. 63
  b. Mostly terrestrial and not pyxidately branched 6
  6a. Usually unbranched,
      spores smooth . . . . *Clavaria*, p. 59
    b. Branched, spores smooth
       or ornamented . . . . *Ramariopsis*, p. 68

Genus *Clavaria*

The majority of the coral fungi were once placed in the genus *Clavaria*. Now it is restricted to species which are primarily terrestrial with white, subglobose to ellipsoid, smooth spores. They have brittle flesh and are mostly unbranched, although a few species are decidedly branched.

Clavaria purpurea Fr.

FRUITING BODY up to 12 cm high; simple and round in cross-section; often thickest midway to tip, then tapering to a point or blunt; purple to brownish-gray tinged with purple; usually growing in dense clumps. SPORES white; ellipsoid; smooth; 5.5-9 x 3-5 μ. EDIBILITY unknown.

C. purpurea is not very common, but it does occur in late summer and fall in western coniferous forests, where this beautiful fungus shows a preference for bare or grassy situations rather than dense leaf litter.

Clavaria vermicularis Fr.

FRUITING BODY usually less than 12 cm high; simple and round in cross-section; tapering to a point; white, becoming yellow at tip with age. FLESH brittle. SPORES white; ellipsoid; smooth; 5-7 x 3-4 μ. EDIBLE.

FIG. 9. *Clavaria vermicularis.*

C. vermicularis (Fig. 9) is found throughout the western states. It occurs commonly in grassy situations in coniferous and broadleaf woods, as well as along the edges of brushy areas in fall and early winter, often growing in dense clumps. The name is derived from its wormlike, twisted shape. Its flesh is tasteless and so delicate that it seems to dissolve in one's mouth.

Genus *Clavariadelphus*

Species of *Clavariadelphus* have simple club-shaped or cylindrical fruiting bodies and range in height from 2.5 to about 20 cm; the apex may be rounded, truncate, or pointed. Spores are produced on most of the exterior surface, which is smooth at first but becomes wrinkled longitudinally at maturity. In *C. truncatus*, *C. borealis*, and *C. pistillaris*, the length of the spores is no more than twice their width, whereas in *C. sachelinensis* and *C. ligula* the length of the spores is at least three times their width. The wrinkles or folds on the fruiting body are considered by some to be rudimentary lamellae.

KEY TO SPECIES OF *CLAVARIADELPHUS*

1a. Spores ochre 2
 b. Spores white 3
  2a. Apex truncate .... *C. truncatus*, p. 61
   b. Apex not truncate .... *C. sachelinensis*, p. 63
3a. Apex truncate .... *C. borealis*, p. 62
 b. Apex not conspicuously truncate 4
  4a. Fruiting body large and
   staining brown .... *C. pistillaris*, p. 62
   b. Fruiting body smaller and not
    staining brown .... *C. ligula*, p. 63

Clavariadelphus truncatus (Quél.) Donk
Clavaria truncatus Quél.

FRUITING BODY up to 15 cm high; club-shaped, with apex noticeably truncate; tapering toward base; surface vertically and irregularly wrinkled; pinkish-brown; usually orange-yellow at apex. FLESH white, sweet-tasting. SPORES ochraceous; broadly ellipsoid; smooth; 9-11.5 x 6-7 μ. EDIBLE and good.

C. truncatus has widespread distribution throughout the temperate zone of the northern hemisphere, and it is associated with coniferous trees. Its flesh has a sweet taste, but it tends to become spongy with age.

C. borealis Wells and Kempton (Plate 18) is very similar in external appearance to *C. truncatus*, but it has white instead of ochraceous spores. It is also associated with conifers and is common from Alaska to Oregon and northern Idaho in late summer and fall. It is edible but is reported to have a somewhat bitter taste. It is possible that *C. borealis* may prove to be more widespread and abundant in the West than *C. truncatus*.

For many years *C. truncatus* and *C. pistillaris* (see below) have been confused and often thought to represent but a single species. *C. pistillaris*, however, differs from *C. truncatus* in having white spores, and differs from both *C. truncatus* and *C. borealis* in being rounded rather than flattened at the apex. It is also associated with deciduous rather than coniferous trees.

All three species are edible and can be eaten raw. The flesh is sweet and brittle.

Clavariadelphus pistillaris (Fr.) Donk
Clavaria pistillaris L. ex Fr.

FRUITING BODY 7.5-20 cm high; club-shaped and tapering downward from rounded apex; surface dull, smooth until maturity, when irregular vertical wrinkles and grooves develop; yellow at apex, becoming yellowish-brown below, white at base. FLESH white, staining vinaceous-brown when cut; taste mild. SPORES white; ellipsoid; smooth; 9-14 x 4.5-7.5 μ. EDIBLE and good.

C. pistillaris is a widely distributed species, common in the West from Alaska southward. It is most frequently found in fall and winter in nonconiferous woods.

Clavariadelphus sachelinensis (Imai) Coker

FRUITING BODY 2.5-7.5 cm high; club-shaped, tapering toward base; apex rounded, acute, even flattened at times; surface irregularly wrinkled longitudinally; yellowish to pallid, sometimes with vinaceous tinge; white; tomentose at base. FLESH white, does not change color when cut. SPORES ochraceous; narrowly ellipsoid; smooth; 16-24 x 4-6 μ. EDIBILITY unknown.

C. sachelinensis is common in western coniferous forests. **C. ligula** (Fr.) Donk, which is found in the same region, is very similar, but may be distinguished by its white spores, which are smaller, measuring 12-15 x 3-4.5 μ. Its edibility is also unknown.

Genus *Clavicorona*

A small group of coral fungi whose branches arise from the margin of the apex of the branch below are placed in the genus *Clavicorona*. Each crownlike apex from which new branches arise becomes flattened or concave, more or less forming a cup. Such a form of branching is called *pyxidate*.

Members of *Clavicorona* occur on rotting wood, needles, or leaves. They have small white spores which may be subglobose or ellipsoid and are either smooth or ornamented.

Clavicorona avellanea Leathers and Smith

FRUITING BODY up to 8 cm high; several tiers of erect branches, those at base somewhat pubescent, terminal ones smooth; pale vinaceous-brown at base, becoming pallid toward tips; color unchanged when cut or bruised. FLESH brittle, acrid and burning to taste. SPORES white; ovoid; smooth to minutely roughened; 3.5-4.5 x 3-4 μ. EDIBLE, but poor.

C. avellanea (Fig. 10) grows on rotten coniferous logs; it is known from northern Idaho and probably ranges over

much of the Pacific Northwest. As in other members of
Clavicorona, the apex of each branch is somewhat cup-
shaped, with marginal, erect tips.

FIG. 10. *Clavicorona avellanea.*

Clavicorona taxophila (Thom) Doty

FRUITING BODY up to 2.5 cm high; unbranched; cylindrical;
tapering downward toward base; apex flat or concave; white
to slightly yellowish, somewhat translucent; a few hairs
present at base. FLESH soft, waxy. SPORES white; subglobose; smooth; 3-4 x 2-3 μ. EDIBILITY unknown.

    C. taxophila appears on wood, needles, or rotting leaves
where, because of its small size, it may easily be overlooked.
This small white coral fungus grows as a single stalk,
flattened or cup-shaped on top with a few small projections
arising from the margin of the cup.

Genus *Clavulina*

Species of *Clavulina* are mostly terrestrial. The fruiting body is usually branched, with 2-spored basidia; spores are white, subglobose, and smooth, and contain a large oil drop.

Clavulina cinerea (Fr.) Schroet.
Clavaria cinerea Fr.

FRUITING BODY up to 10 cm high; repeatedly branched; branches longitudinally ridged and usually blunt at tips; dark purplish-gray, whitish at base. SPORES white, nearly round; smooth; 6.5 x 11 μ. EDIBLE.

C. cinerea is terrestrial, growing most frequently under conifers in late summer and fall. It has widespread distribution in the western states and provinces. Because of its fair size and grayish color, this is an attractive species.

C. cinerea might be confused with **Ramaria fennica** var. **violaceibrunnea** Marr and Stuntz, which is also grayish-purple or violet with a white base. *R. fennica* var. *violaceibrunnea*, however, has yellowish, ellipsoid, warted spores. Upon application of 20-percent KOH to the pigmented surface of the flesh it quickly turns an orange-red, whereas *C. cinerea* does not.

Clavulina cristata (Fr.) Schroet.
Clavaria cristata Fr.

FRUITING BODY up to 8 cm high; sparingly branched, with tips *cristate* (having small pointed projections); white or cream-color. SPORES white; subglobose; smooth; 7-11 x 6-10 μ. EDIBLE, but unpalatable.

C. cristata is very common in both coniferous and broadleaf forests in the West. Sometimes the ground seems literally covered with this small white coral fungus. It is found from late summer to winter. Although non-poisonous, its tough flesh and lack of substance make it unappealing to the mushroom eater.

Clavulina rugosa (Fr.) Schroet.
Clavaria rugosa Fr.

FRUITING BODY up to 10 cm high; simple or with a few short branches which are longitudinally ridged and have blunt tips; white to cream-color. SPORES white; subglobose; smooth; 9-14 x 8-12 μ. EDIBLE.

C. rugosa is a widespread terrestrial species which may occur in coniferous or nonconiferous woods or even in open fields. Upon drying, the specimen becomes yellow to orange.

C. rugosa resembles *C. cristata* (see above), but differs in being less branched, with the branches more ridged and lacking cristate tips. The spores of *C. rugosa* are also larger.

Genus *Ramaria*

Ramaria is the genus in the family Clavariaceae with the most species. A few species are small, but the majority are large and conspicuous. All are radially branched and show no flattening of the apex prior to the next branching. The spores are pale yellow to rusty-cinnamon, ellipsoid, smooth or ornamented.

Ramaria is a difficult genus, one that gives both the amateur and professional trouble in identification. Some common western species are described here, but for many others it will be necessary to consult special monographs listed in the Bibliography.

Ramaria botrytis (Pers. ex Fr.) Rick.
Clavaria botrytis Pers.

FRUITING BODY up to 20 cm long; arising from a thick base; branches stout and numerous, with tips crowded together; base white, branches cream-color to tan, with pink to purplish-red tips. SPORES golden-yellow; ellipsoid; nearly smooth; 12-16 x 4-6 μ. EDIBLE.

R. botrytis occurs from late spring to fall in both coniferous and nonconiferous forests of the Pacific Coast and

Rocky Mountain regions. It is a robust terrestrial species with a cauliflower-like appearance. The size, pallid color of the branches, with pink to reddish tips, and flesh that does not change when bruised are good identifying characters.

A somewhat similar species, tentatively referred to here as **R. flava** var. **sanguinea** (Coker) Corner, differs in having yellowish tips to the branches and flesh that rapidly bruises wine-color, especially near the base. It is fairly common under conifers in the higher Sierra Nevada in fall.

Ramaria flavobrunnescens (Atk.) Corner
Clavaria flavobrunnescens Atk.

FRUITING BODY large, up to 20 cm high; branching up to six times, with branches thick; base short, thick, white; branches yellow when young, becoming tan with age. FLESH white, does not stain where bruised; taste not distinctive; odor sometimes slightly sweetish. SPORES yellow to light ochraceous; subcylindrical; minutely roughened; 9-12 x 3-5 μ. EDIBLE.

R. flavobrunnescens is a common large coral fungus that occurs in the fall and winter along the Pacific Coast and in late spring and fall in the Rocky Mountains. The cauliflower-like fruiting bodies push up through conifer-needle litter. It does not stain on bruising, and lacks any bitter taste.

R. flavobrunnescens var. **aromatica** Marr and Stuntz, described from western Washington where it occurs under Western Hemlock, has a distinctly sweet odor. Its edibility is unknown.

Ramaria formosa (Fr.) Quél.   Plate 19
Clavaria formosa Fr.

FRUITING BODY up to 20 cm high, with massive base; branching up to six times from base, with lower branches thick, tips rounded or pointed; base white to brownish-white; branches pinkish or salmon-color to yellowish-tan in age, tips yellow when young. FLESH bruises brownish. SPORES dull yellow; ellipsoid; ornamented with large warts; 9-12 x 4.5-6 μ. POISONOUS.

R. formosa is a widespread terrestrial species that is most common in the West under Douglas Fir and hemlocks, but is not limited to such conifers. *R. formosa* is poisonous, and since there are several other species of somewhat similar external appearance, do not eat large tan or salmon-pink coral fungi!

R. testaceoflava var. **brunnea** (Zeller) Marr and Stuntz also bruises brown and has branches which vary from yellow to brown; its flesh is distinctly bitter and thus inedible.

Ramaria stricta (Fr.) Quél.   Plate 20
Clavaria stricta Fr.

FRUITING BODY up to 14 cm high; branched repeatedly, with branches slender, vertical, and compact; light tan to vinaceous-brown. STIPE short or absent. SPORES cinnamon-buff; ellipsoid; minutely ornamented; 7-10 x 4-5 μ. EDIBLE but unpalatable.

R. stricta is lignicolous, common in late summer and fall in coniferous forests of the Pacific Coast and Rocky Mountains. Its slender vertical branches are distinctive.

R. apiculata (Fr.) Donk is somewhat similar but has green pigmentation. Its edibility is unknown.

Genus *Ramariopsis*

Ramariopsis consists of terrestrial, branched coral fungi with white, ornamented spores ranging from subglobose to ellipsoid in shape. There is only one species now known in the West.

Ramariopsis kunzei (Fr.) Donk
Clavaria kunzei Fr.

FRUITING BODY up to 12 cm high; usually much branched, with the axils rounded, branches erect, and tips often pointed; white with base sometimes tinged pink. SPORES white; globose; minutely ornamented; 3.5 x 4.5 μ. EDIBILITY unknown.

R. kunzei occurs most commonly in deciduous woodlands, but may be found in coniferous forests along the Pacific Coast from California north. This widespread delicate white coral fungus is primarily terrestrial, but does grow also on very rotten wood.

Genus *Sparassis*

Sparassis is distinguished from all other genera of coral fungi by its broad, flattened branches. It has aptly been described by Dr. A. H. Smith as resembling "a bouquet of egg-noodles."

Sparassis radicata Weir   Plate 21

FRUITING BODY up to 35 cm wide and 50 cm high, arising from a thick, fleshy base; branches flattened and wavy; creamy to yellow. FLESH white. SPORES white; ellipsoid; smooth; 5-6 x 3-3.5 μ. EDIBLE and choice.

S. radicata is found under conifers in the fall and winter in the western states and provinces. In coastal California it is often associated with Bishop Pine (*Pinus muricata*). This is a fine species for the table.

CHANTERELLES

FAMILY CANTHARELLACEAE

The fruiting body in the Cantharellaceae varies from a regular pileate and stipitate form to tubular or funnel-shaped. The spore-bearing surface is smooth, veined or with shallow gill-line blunt ridges, fleshy or membranous. Spores are white to ochre, variable in shape.

KEY TO COMMON GENERA
OF CANTHARELLACEAE

1a. Spores ochre, usually ornamented .. *Gomphus*, p. 70
  b. Spores white to pale pinkish-buff 2
    2a. Fruiting body usually yellow
        to orange-red .... *Cantharellus*, p. 72
     b. Fruiting body blackish
        or gray .... *Craterellus*, p. 73

Genus *Gomphus*

The fruiting body in *Gomphus* is large, fleshy, trumpet-shaped to club-shaped; spores are yellowish-orange, usually tuberculate or wrinkled.

Gomphus clavatus (Fr.) S.F. Gray   Plate 22
Cantharellus clavatus Fr.

PILEUS up to 20 cm across; often club-shaped when young, but expanding greatly to become plane, then depressed centrally, sometimes funnel-shaped; margin irregularly lobed, often higher on one side; surface smooth; dry; yellowish-brown, often with violet tinge. FLESH soft; white. GILL-LIKE RIDGES shallow, blunt, and veinlike; close; interconnected; purple or lavender, becoming grayish with age. STIPE up to 8 cm long; blending with pileus; often compound, branching into two or more fruiting bodies. SPORES orange-yellow; ellipsoid; wrinkled; 10-15 x 4-6 μ. EDIBLE and choice.

G. clavatus is common in coniferous forests in the West. It fruits principally in fall and early winter and shows a preference for deep litter in moist, shady situations. The deep purple, veined gill-like ridges and large size are good characters for identification. This is an edible, choice mushroom when young and, as it is commonly caespitose, one clump will provide a nice addition to a meal. The flesh becomes bitter with age.

Gomphus floccosus (Schw.) Sing.   Plate 23
Cantharellus floccosus Schw.

PILEUS up to 15 cm across; depressed centrally; trumpet-shaped; surface coarsely scaly; dry; orange-yellow to reddish-orange. FLESH white; lacking a distinct odor or taste. GILL-LIKE RIDGES consist of irregular veins extending down stipe, which is not distinct from pileus; white to yellowish. SPORES ochre; ellipsoid; wrinkled; 12-15 x 6-7.5 μ. EDIBILITY questionable; see below.

G. floccosus is common in coniferous forests of the Pacific Coast in autumn. While some people seem able to eat it with no trouble, it is known to contain gastrointestinal irritants and has caused illness requiring hospitalization.

Two other closely related species of *Gomphus* have been

described from western North America: **G. bonari** (Morse) Sing. is said to be lemon-yellow at the base; **G. kauffmanii** (A. H. Smith) Corner has a large pileus, clay-color to tawny, with very large scales on the surface. *G. kauffmanii* is fairly common in higher parts of the Sierra Nevada, where it occurs in clumps in early autumn. Like *G. floccosus*, it is suspect, and none of these three species should be eaten.

Genus *Cantharellus*

The fruiting body in *Cantharellus* is fleshy, with pileus and stipe; gill-like ridges are distinct, although often shallow and interveined, and spores are white to yellow or pinkish; it is terrestrial.

Cantharellus cibarius Fr.   Plate 24

PILEUS up to 10 cm across; broadly convex, but soon becoming centrally depressed; margin irregular and wavy; smooth; dry; pale to orange-yellow. FLESH light yellow; firm; often with a fruity odor vaguely resembling apricots. GILL-LIKE RIDGES shallow; blunt; forked; intervenose; extending down stipe; usually paler than surface of pileus. STIPE up to 8 cm long; smooth; similar in color to pileus. SPORES yellow; ellipsoid; smooth; 8-11 x 4-5.5 μ. EDIBLE and choice.

C. cibarius, the Golden Chanterelle, occurs in hardwood and conifer forests throughout the temperate zone of the northern hemisphere. It is one of the best known and most popular of all mushrooms. A 1975 questionnaire sent to about 400 members of the Mycological Society of San Francisco listed it most often in the ten favorite mushrooms (65.2%). It usually grows under oaks but is often hidden under fallen leaves. Since its color blends well with that of the leaves, you must search for it in the fall and winter. Once found in association with certain trees, it can often be collected in the same location year after year. Its blunt ridges are so distinctive that it would be hard to confuse it with a similarly colored gilled mushroom, but mistakes have been made (see *Clitocybe aurantiaca* and *Omphalotus olivascens*).

A closely related species, **C. subalbidus** Smith and Morse, called the White Chanterelle, occurs along the Pacific Coast from central California northward. In shape and habitat it is very similar to *C. cibarius*, but it has a white fruiting body that stains orange where bruised, and its spores are white and larger than those of *C. cibarius*. Both may be found in the same general area and are equally edible.

Genus *Craterellus*

The fruiting body in *Craterellus* is black or grayish, funnel-shaped, with thin flesh. The spore-bearing surface is smooth or only slightly wrinkled; spores are white to salmon-buff.

Craterellus cornucopioides Fr.

FRUITING BODY up to 6 cm in diameter and 8 cm high; trumpet-shaped; deeply depressed centrally; blackish. FLESH very thin. SPORE-BEARING SURFACE wrinkled or nearly smooth; ashy-gray. STIPE hardly distinguishable. SPORES white; ellipsoid; smooth; 11-15 x 7-9 μ. EDIBLE and choice.

C. cornucopioides grows in clusters in coniferous and mixed forests in the West in winter. Because of its dark color, it is easily overlooked on the humus of the forest floor. However, it is well worth searching for. In our opinion it is a most delicious mushroom, although its small size and thin flesh necessitate gathering a great many fruiting bodies for a substantial dish.

WOODY TUBE FUNGI/Polypores

FAMILY POLYPORACEAE

The Polyporaceae, commonly called polypores, are usually woody, leathery, or membranous, although a few are fleshy. The spores are generally produced within pores on the undersurface of the fruiting body, although a few kinds have lamellae-like structures. Some species are stipitate, but the majority are bracket-shaped or shelf-like, growing on dead or living wood. The spores are variable in shape; they may be smooth or ornamented, and range from white to yellow or brown. Some polypores are serious parasites on forest trees, but the majority are saprophytes on snags, stumps, and decaying logs.

KEY TO COMMON GENERA
OF POLYPORACEAE

1a. Fruiting body with a crustlike upper surface;
    spores brown, warted, truncate
    at one end .... *Ganoderma*, p. 75
 b. Not as above 2
    2a. Fruiting body woody, perennial ... *Fomes*, p. 77
     b. Not as above 3
3a. Fruiting body stipitate 4
 b. Fruiting body not stipitate 5
    4a. Growing on wood .... *Polyporus*, p. 78
     b. Growing on the ground .... *Coltricia*, p. 79
5a. Spore-bearing surface covered ... *Cryptoporus*, p. 80
 b. Spore-bearing surface not covered 6
    6a. Spore-bearing surface lamellate .. *Lenzites*, p. 81
     b. Spore-bearing surface poroid 7
7a. Fruiting body fleshy,
    pore surface yellow.............. *Laetiporus*, p. 82
 b. Fruiting body thin, leathery,
    colorfully zoned .... *Coriolus*, p. 83

Genus *Ganoderma*

Ganoderma is one of the two genera that contain most of the large, perennial, woody bracket or shelf fungi, also called "conks." Most of them grow in a fanlike or hoof-shaped form on the trunks of living or dead trees. The crustlike upper surface, which is often shiny, and the brown, warted spores, truncate at one end, usually distinguish species of *Ganoderma* from those of *Fomes* (see below).

Ganoderma applanatum (Pers. ex Wallr.) Pat.
Fomes applanatus (Pers. ex Wallr.) Gill.

FRUITING BODY up to 50 cm in diameter and 10 cm deep; fan-shaped; surface has a crust and, usually, irregular nodules; grayish, often indistinctly zonate. PORE SURFACE white when young, instantly bruises darker; tubes distinctly stratified, each season's growth recognizably separable. SPORES brown; ovoid, with one end truncate; minute spines; 6-9 x 4.5-6 μ. INEDIBLE.

G. applantum (Fig. 11) is probably the most widespread bracket or shelf fungus in North America. It occurs most frequently on broadleafed trees but is also found on conifers throughout the United States and Canada. The large fan shape and gray color of the upper surface, and the instant staining of the pore surface when incised, are good field characters. Because one can draw or write on the underside of this species, it has been called the Artist's Fungus.

FIG. 11. *Ganoderma applanatum.*

Ganoderma oregonense Murr. Plate 25
Polyporus oregonensis (Murr.) Kauf.

FRUITING BODY up to 80 cm in diameter; shelf-like or *reniform* (kidney-shaped), usually with an eccentric stipe; upper surface shiny, red, or mahogany-colored; sometimes *sulcate* (grooved) on the margin. PORE SURFACE white, but discolors when handled, becoming brown with age. SPORES brown; ovoid with one end truncate; *echinulate* (spiny); 10-16 x 7.5-9 μ. INEDIBLE.

G. oregonense can be found on the dead trunks or stumps of coniferous trees from Alaska south to California, and also in the Rocky Mountains. The large size, the shiny red surface of the fruiting body, and the usual presence of a stalk are striking characters.

G. oregonense very much resembles **G. tsugae** Murr. of eastern North America, but has larger spores. **G. lucidum** (Fr.) Karst. is also similar to *G. oregonense*, but it is less shiny on the upper surface and grows on living deciduous trees and stumps. It too is inedible.

Genus *Fomes*

Fomes contains species of perennial woody pore fungi that are typically *ungulate* (hoof-shaped). Each season new growth is added to the margin and there is a downward extension of the hymenial area. The upper surface is hard and often *zonate* (marked with concentric bands of color), although it is not regarded as having a true crust as in the genus *Ganoderma* (see above). This, however, is often difficult to determine, especially in species of *Fomes* where there is a resinous coating on the upper surface that subsequently hardens.

The principal difference between the two genera is in spore shape. The spores in *Ganoderma* are brown, warty, and truncate on one end, and those of *Fomes* are not. Since this distinction is relatively slight, some authors do not recognize *Ganoderma* as a separate genus.

Fomes pinicola (Fr.) Cke.

FRUITING BODY up to 40 cm in diameter; typically hoof-shaped; surface usually zoned and partly covered by a resinous crust that varies from reddish-brown to black, with outer margin white. PORE SURFACE whitish. SPORES colorless; ovoid; smooth; 5-7 x 4-5 μ. INEDIBLE.

F. pinicola, variously known as the Pine Destroyer and the Red Belt Fungus, is common throughout both the coniferous and hardwood forests of western United States and Canada. It is not an important destroyer of living trees, being found mostly on stumps and standing snags. It can be recognized by the black and red resinous zones and the white margin.

The related **F. annosus** (Fr.) Cke., common in Western Yellow and Jeffrey Pine forests in the West, is a very destructive species. It commonly grows at the base of the trunk of these and other conifers as well as a few hardwood species, and is even found on mine timbers. The rot that it produces is known as "spongy sap rot." Unlike *F. pinicola*, which is ungulate, the fruiting body of *F. annosus*, which may be 15 cm in diameter, is flattened over the trunk, with the margin turned back to form a pileus. It is grayish-brown to dark brown, sometimes with a tinge of red. When young it is somewhat woolly, but becomes smooth with age.

Genus *Polyporus*

The genus *Polyporus* is now generally restricted to stipitate polypores that grow on wood, while formerly in the broad sense used by Overholts (1953), it included the majority of the Polyporaceae. The spores are white and smooth. Two common species are described here.

Polyporus elegans Bull. ex Fr.

PILEUS up to 7 cm in diameter; round to kidney-shaped; pale tan, becoming whitish with age; smooth or with a powdery bloom. FLESH white; leathery, becoming rigid when dry. PORE SURFACE grayish to pallid; tubes decurrent on stipe; tube mouths angular. STIPE up to 5 cm long; slender; central or eccentric; base or lower half black. SPORES white; cylindric; smooth; 6-10 x 2.5-3.5 μ. INEDIBLE.

P. elegans is a very common small, stipitate polypore, usually occurring in mountainous areas where there are mixed forests from Alaska and the Yukon Territory southward. It is most often found on fallen branches of deciduous trees, although it occasionally grows on dead branches of conifers. The pale color of the pileus and the black base of the stipe are good field characters.

Polyporus picipes Fr.

PILEUS up to 20 cm in diameter (usually much smaller); round; convex when young, becoming depressed with age, sometimes funnel-shaped; margin often lobed; surface reddish-brown, becoming blackish with age. FLESH whitish; leathery; rigid on drying. PORE SURFACE white at first, later becoming pale brown; tubes somewhat decurrent on stipe; tube mouths very small, circular to angular. STIPE up to 6 cm long; relatively thin; central or eccentric; black, at least on lower half. SPORES white; cylindric to ellipsoid; smooth; 6-8 x 3-4 μ. INEDIBLE.

P. picipes, like *P. elegans*, is common in the West, from Alaska south to California and Idaho where it most frequently grows on stumps and rotting logs. The rich, chestnut-red pileus and the minute (almost invisible to the naked eye) pores readily distinguish it from *P. elegans* (see above).

Genus *Coltricia*

Coltricia includes a few species of terrestrial, stipitate polypores that are very similar in most characters to members of the genus *Polyporus* (see above), but have a leathery, rusty-brown, generally zonate pileus and flesh that is also rusty-brown. The stipe is well developed and central. The hymenial surface is poroid in most species, but may break up into concentrically arranged lamellae around the stipe. The spores vary from colorless to brown, are globose to cylindrical or allantoid, and smooth.

Coltricia perennis (L. ex Fr.) Murr.
Polyporus perennis L. ex Fr.

PILEUS up to 11 cm in diameter; round to irregular; often depressed; surface velvety; concentrically zonate; rusty-brown. FLESH thin; leathery; brown. PORE SURFACE yellow brown to cinnamon or gray; tubes somewhat decurrent; tube mouths angular. STIPE up to 7 cm long; central; velvety-brown. SPORES white; oblong-ellipsoid; smooth; 6-9 x 3-5 μ. INEDIBLE.

C. perennis, which ranges from Alaska south to California, Arizona and Colorado in the West, is principally terrestrial, growing along trails or in burned areas in the woods. Occasionally it grows on rotting wood. It is an attractive fungus with its soft, velvety, zonate, rusty-brown pileus.

Genus *Cryptoporus*

The name *Cryptoporus* is based on the presence of a membranous outgrowth of the margin of the fruiting body that encloses and hides the spore-bearing surface.

Cryptoporus volvatus (Pk.) Hubbard
Polyporus volvatus Pk.

FRUITING BODY up to 8 cm in diameter; sessile; globose to oval; cream-color to golden-brown. FLESH white; fibrous. PORE SURFACE yellowish-brown; covered by a layer, continuous with the margin of the fruiting body, which is eventually perforated by holes made by insects. SPORES flesh-colored; elongate-ellipsoid to short cylindric; smooth; 8-12 x 3-5 μ. INEDIBLE.

C. volvatus, a cream-color to golden-brown egg-shaped fungus with its pore surface covered by a thin membrane, cannot be confused with any other. It occurs principally on dead conifers, standing or fallen, and occasionally on living trees. The spores escape through a hole or holes produced by insect larvae in the hymenial covering. In western North American it occurs from southern Canada south to California, Nevada, Utah and Colorado.

Genus *Lenzites*

The fruiting body in *Lenzites* is flat and horizontally expanded into a fan-shaped semicircle. The flesh is rather thin and leathery to corky. The pore-bearing surface on the underside usually consists of lamellae-like plates which radiate from the area of attachment.

A closely related genus, **Daedalea**, can easily be confused with *Lenzites*; in fact, all the species now in *Lenzites* were at one time placed in *Daedalea*. Most species of *Daedalea* can be separated from those of *Lenzites* by the poroid hymenium: the walls of the pores are thick, and the mouths may be elongated so much that they appear lamellate. Sometimes they are toothed, as in **D. unicolor** Bull. ex Fr., a common species on the dead wood of deciduous trees in the West.

Lenzites betulina (L. ex Fr.) Fr.   Plate 26
Daedalea betulina L. ex Fr.

FRUITING BODY up to 12 cm in diameter; leathery and rather flexible; upper surface tomentose; grayish or grayish-brown with multicolored zones, often with a green algal growth. FLESH white; leathery to corky. PORE SURFACE lamellate, with occasional pores; whitish. SPORES white in mass; cylindric; smooth; 4-7 x 1.5-3 μ. INEDIBLE.

L. betulina is a very common species likely to be encountered by most collectors; it is found on dead wood, principally of deciduous trees, but also occasionally on dead conifers, throughout the western states and north to British Columbia and Alberta. Although it has lamellae with only occasional pores, it belongs in the Polyporaceae. The upper surface is very distinctive, with its tomentose, multizoned coloration somewhat like that of some fruiting bodies of *Coriolus versicolor* (see below). The green algal growth appears as it ages.

Genus *Laetiporus*

Laetiporus contains the sulphur mushrooms, large, colorful, fleshy shelf fungi with a bright yellow pore surface. Sulphur mushrooms, so-called because of the color of the hymenial surface, grow in large shelflike masses on logs, stumps, and living trees, both coniferous and broadleafed.

Laetiporus sulphureus (Bull. ex Fr.) Murr.
Polyporus sulphureus Bull. ex Fr.

FRUITING BODY up to 60 cm broad; sessile or substipitate; usually in imbricate clusters or rosettes; upper surface bright orange with yellow margin when young, becoming grayish-white with age; margin wavy. FLESH firm but tender when young; white to yellowish. PORE SURFACE bright sulphur-yellow when young, fading with age; tube mouths angular, sometimes slightly toothed. SPORES white; ellipsoid to ovoid; smooth; 5-7 x 3.5-4.5 μ. EDIBILITY questionable; see below.

L. sulphureus, a very widespread species, is found in late summer in the West from Alaska south. Altitudinally it occurs from sea level to over 9,000 feet. The margin of the young fruiting body is regarded as edible and tastes somewhat like chicken when fried; the name Chicken Mushroom is sometimes applied to it for this reason. However, several members of the Mycological Society of San Francisco have reported sudden and violent reactions to this mushroom, and it seems likely that there is now a toxic strain of *L. sulphureus* on the Pacific Coast.

Genus *Coriolus*

The fruiting body in *Coriolus* is leathery and sessile, in the form of a smooth, thin shelf with the upper surface markedly zonate. Fruiting bodies usually occur in numbers, either overlapping one another or in the form of a rosette.

Coriolus versicolor (L. ex Fr.) Quél.
Polyporus versicolor L. ex Fr.

FRUITING BODY up to 10 cm broad; sessile; multilayered or in form of rosette; upper surface silky or velvety; numerous narrow, multicolored zones of brown, gray, yellow, and red, or blue, gray, and black. FLESH white; leathery. PORE SURFACE white to yellow; tube mouths angular, small but visible to unaided eye. SPORES white; cylindrical or allantoid; smooth; 4-6 x 1.5-2 μ. EDIBLE.

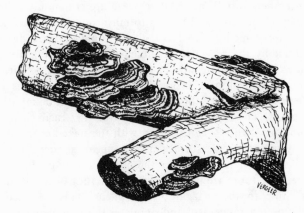

FIG. 12. *Coriolus versicolor.*

C. versicolor (Fig. 12) is found in broadleafed and coniferous forests nearly everywhere in North America. It usually
grows on dead wood in the late summer and fall. The
numerous overlapping small shelves or rosettes with their
multicolored zones make members of this species an attractive part of the forest environment.

FLESHY TUBE FUNGI/Boletes

FAMILY BOLETACEAE

Members of the family Boletaceae are commonly called
"boletes," and are well known to mushroom hunters. They
are mostly medium to large, fleshy fungi that are distinguished by the presence of united tubes extending downward from the underside of the pileus. The spore-bearing
surface (hymenium) is inside these tubes, and the spores are
released at maturity through openings called tube mouths,
or pores. These pores may be so large that their shape—
round or angular—can be seen with the naked eye; but in
many cases the undersurface merely appears spongelike, and
a hand lens is needed to see the minute openings. In some
genera the tubes are formed in rows radiating outward from
the stipe to the margin of the pileus. This is called a
boletinoid arrangement and simulates the alignment of gills
in the agarics (see next chapter).

The spores of boletes are most often yellowish-brown or olive-brown, but in a few cases they may be yellow, pink, reddish-brown, or blackish. They are usually ellipsoid to subcylindric and smooth, but they may be subglobose and ridged, echinulate, or reticulate.

Boletes are usually terrestrial; in a few cases they are lignicolous or are parasitic on another fungus. They are mostly associated with conifers, and some are definitely restricted to the vicinity of specific trees such as Western White Pine or species of larch.

Almost all of the boletes were formerly lumped into a single genus, *Boletus*, and classified in the family Polyporaceae along with the woody tube fungi. Modern taxonomy has now placed them in a separate family and divided *Boletus* into many different genera.

Many boletes are edible and choice for the table, but some of the glutinous or viscid species are rather slimy if the pellicle is not removed before cooking. A few species are known to be poisonous—those to be especially avoided are the ones with red tube mouths.

KEY TO GENERA OF THE BOLETACEAE

1a. Stipe with white scales at first, which become
brown or black on tips . . . . *Leccinum*, p. 86
  b. Stipe smooth, glandular, or reticulate 2
    2a. Spores pale cinnamon, yellow-brown, olive-
brown, or dark brown 3
      b. Spores light reddish-brown, purplish-
brown, or rusty 4
3a. Veil dry, bright yellow . . . . *Pulveroboletus*, p. 100
  b. Veil often present; pileus usually viscid;
stipe often glandular . . . . *Suillus*, p. 87
  c. Veil absent; stipe smooth, scaly,
or reticulate . . . . *Boletus*, p. 100
    4a. Pores angular, radially arranged;
veil present . . . . *Fuscoboletinus*, p. 97
      b. Pores round; veil absent . . . . *Tylopilus*, p. 99

Genus *Leccinum*

Members of *Leccinum* are typically large boletes with a cap that is convex when young and generally remains so. The margin of the pileus is usually noticeably membranous and extends beyond the tubes. The surface of the pileus may be dry or viscid and varies from smooth to granular or tomentose; in most western species it is orange, brick-red, or brownish-black. The tubes have very small pores and are usually white when young, becoming grayish-brown or almost black with age. The stipe is solid; it may be equal except near the pileus, where it is often constricted, or may gradually widen from apex to base. An important generic character is the ornamentation in the form of short projections on the stipe that darken with age. Spores are generally dark yellow to olive-brown. Members of this genus are considered to be very choice eating.

The generic characters are distinct and easily recognizable, but the large number of closely related forms of *Leccinum* recently described by Smith and Thiers make specific identification difficult.

Leccinum aurantiacum (Bull. ex St. Amans) S.F. Gray

PILEUS up to 10-30 cm in diameter; convex, ultimately becoming almost plane; surface dry or slightly viscid, and rough; margin extends beyond tubes as sterile membrane; bright rust brown to brick red. FLESH solid; white; slowly changes to vinaceous when cut. TUBES grayish; small and angular; depressed near stipe. STIPE 10-16 cm long, about 3 cm in diameter; equal or constricted at apex, at least when young; surface white but ornamented with small points, which may be light at first but darken with age to blackish; flesh white but, when cut, slowly turning vinaceous then darkening, sometimes becoming deep blue or reddish at base. SPORES dark yellow to dull cinnamon; elongate and tapering; smooth; 12-15 x 4-5 μ. EDIBLE and choice.

L. aurantiacum is widely distributed, particularly where

there are aspens and pines. In parts of Alaska it grows on the tundra.

L. insigne Smith, Thiers, and Watling is a very common bolete often mistaken for *L. aurantiacum*. However, it is associated only with aspens, its pileus is some shade of reddish-brown, and the flesh slowly changes to a dark smoky brown. **L. ponderosum** Smith, Thiers, and Watling, which was described from southern Oregon, is similar in the color of the pileus but may attain a very large size, with specimens up to 30 cm in diameter being recorded. The stipe turns blue when handled, and the flesh of the pileus does not become gray or vinaceous when cut. In northern Idaho it occur with **L. subtestaceum** Smith, Thiers, and Watling, which has a dull-reddish pileus, tube mouths which are cinnamon-brown rather than gray when young, and a stipe which, when cut, does not stain vinaceous before darkening.

L. manzanitae Thiers is a common species in northern California, occurring under manzanitas and Madrone. It has a large, dark red, viscid pileus similar to that of *L. ponderosum*, but the surface is pitted or reticulate instead of smooth, and the flesh darkens when exposed. **L. scabrum** (Bull. ex Fr.) S.F. Gray (Plate 27) occurs from northern Idaho to Alaska and is usually associated with birches, even the Dwarf Birch of the tundra. It has a dull, grayish-brown pileus which may be dry or viscid, pallid tubes with small pores which become brown with age and may or may not stain yellow when bruised, and a stipe with blackish ornamentation and white flesh which may slowly stain pink. All species of *Leccinum* are considered edible.

Genus *Suillus*

Suillus is an important genus for the mushroom hunter, since it contains many species, a number of which are collected for food. In their study of this group, Smith and Thiers (1964) recognized over 50 species and varieties occurring in North America alone. Some are found only in the East and South, but the Pacific Coast and Rocky Mountains are very well represented. A few species are associated with oaks, Quaking Aspen, and other non-coniferous trees, but the great majority grow under conifers, with which they form a mycorrhizal association.

The pileus in many members of *Suillus* is slimy or viscid. In a few species it is dry, and in some there is a gelatinous layer beneath a relatively dry, scaly outer layer. The pileus is most often some shade of yellow or reddish-brown. The flesh ranges from white to yellow, and in some species changes to blue or blue-green when cut. The tubes are usually yellow but may be white or pallid, as in *S. pungens* when young, to vinaceous-tan, as in *S. tomentosus*; their mouths may be angular, round, or elongate, and in some species are arranged radially. When bruised, the tubes may change to blue, blue-green, or brown. In a number of species the stipe is glandular-dotted; good examples of this are *S. granulatus*, *S. tomentosus*, and *S. pungens*. Remnants of the veil may form a gelatinous annulus, as in *S. subolivaceus*, or a fibrose annulus, as in *S. caerulescens*. Sometimes the remains of the veil adhere to the margin of the pileus. The spores in all western species of *Suillus* are elliptic or subcylindric to almost spindle-shaped and vary from pale to dark yellow-brown.

We have included here those species which we have found most frequently in the West, and which we think the average mushroom hunter is most likely to encounter. None of the species listed is known to be poisonous, and many of them are eaten locally. We suggest removal of the cuticle of viscid forms when preparing them for the table. Like most boletes, the various species of *Suillus* are best sliced and dried for use in sauces.

KEY TO SPECIES OF *SUILLUS*

1a. Stipe hollow .... *S. cavipes*, p. 91

b. Stipe solid 2

    2a. Surface of pileus scaly
and dry .... *S. lakei*, p. 93

    b. Surface of pileus viscid or glutinous 3

3a. Stipe lacks glandular spots when young 4

b. Stipe with glandular spots at all stages 6

    4a. Stipe lacks an annulus .... *S. brevipes*, p. 90

    b. Stipe with floccose or fibrillose annulus 5

5a. Base of stipe turns
blue when cut .... *S. caerulescens*, p. 90

b. Base of stipe does not turn
blue when cut .... *S. grevillei*, p. 92

    6a. Tube mouths vinaceous-tan
when young............. *S. tomentosus*, p. 96

    b. Tube mouths pallid to yellow............... 7

7a. Stipe with glutinous
annulus .... *S. subolivaceus*, p. 95

b. Annulus, if present, not glutinous............... 8

    8a. Droplets of latex present on tubes
when young; odor strong..... *S. pungens*, p. 94

    b. Droplets of latex lacking; odor mild.......... 9

9a. Conspicuous brown spots
on pileus .... *S. sibiricus*, p. 94

b. Pileus umbonate and lacking
brown spots .... *S. umbonatus*, p. 97

c. Pileus reddish-brown or streaked with
reddish-brown on a pale
buff background .... *S. granulatus*, p. 92

Suillus brevipes (Pk.) Kuntze

PILEUS up to 10 cm in diameter; convex, sometimes with margin lobed; surface smooth and glutinous; usually some shade of vinaceous-brown, becoming paler with age. FLESH white at first, becoming yellow with age; does not change color when cut; odor and taste mild. TUBES pallid at first, becoming pale yellow at maturity; mouths small and somewhat angular. STIPE short, rarely over 5 cm long and up to 4 cm thick; white or pale yellow; essentially lacks glandular dots. SPORES cinnamon; elliptic or oblong; smooth; 7-10 x 3 μ. EDIBLE and choice; see genus description above.

S. brevipes sometimes occurs with *S. pungens* under Monterey Pines along the central California coast. In the Pacific Northwest as well as the Sierra Nevada and Rocky Mountains it is associated with Lodgepole Pine. Its firm white flesh is sought by many collectors.

S. brevipes is most likely to be confused with *S. pungens* and *S. granulatus* (see below for both). It can be distinguished from both by the almost complete absence of glandular dots on the short stipe. It further differs from *S. granulatus* in that the stipe does not become bright yellow with age. The odor of *S. brevipes* is mild, unlike that of *S. pungens*.

Suillus caerulescens Smith and Thiers

PILEUS up to 16 or more cm in diameter; broadly convex to plane or centrally depressed; surface with small scales or fibrils; somewhat viscid, or at least moist; tawny-orange. FLESH white or pale yellow, sometimes stains pinkish; odor and taste generally mild. TUBES yellow, staining vinaceous or pinkish-brown. PORES large; neither noticeably angular nor radially arranged. STIPE relatively short and thick in most fruiting bodies, 3-8 cm long and 3 cm thick; surface yellow, but stains brown; lacks glandular dots, but usually somewhat reticulate; annulus fibrillose; flesh at base of stipe turns blue when cut. SPORES dull cinnamon; elliptic; smooth; 8-11 x 3-5 μ. EDIBLE but poor; see below.

S. caerulescens is very common under Douglas Fir along the central California coast, often growing in large aggregations. In the Pacific Northwest and Rocky Mountains it also appears to be associated with Douglas Fir. The tawny-orange or ochraceous-orange pileus with scattered patches of fibrils, combined with a fibrillose annulus and a stipe that turns blue at the base when cut, are identifying characters of this species. The somewhat soft nature of the flesh, the lack of flavor, and the propensity of this species to be infected with fly larvae make it one of the less desirable boletes for the table.

A closely related species, **S. ponderosus** Smith and Thiers, which occurs along the Pacific Coast, is distinguished from *S. caerulescens* by its yellower pileus that lacks fibrils, and by larger pores, a longer annulate stipe, and gelatinous veil. *S. ponderosus* is edible.

Suillus cavipes (Opat.) Smith and Thiers   Plate 28

PILEUS up to 10 cm in diameter; broadly convex, sometimes with umbo, or plane or centrally depressed; surface very tomentose to scaly; dry or moist; yellowish-brown to reddish-brown, with narrow whitish or pallid marginal rim. FLESH white to very pale yellow; does not change when cut; odor and taste mild. TUBES pale yellow at first, becoming deep yellow with age. PORES angular and somewhat elongate; radially arranged and decurrent on stipe. STIPE up to 9 cm long and 1-2 cm in diameter; hollow and sometimes slightly enlarged basally; color somewhat like pileus; white or pale yellow flesh that does not change color when cut; annulus rarely evident. SPORES olive-brown; ovate; smooth; 7-10 x 3.5-4 μ. EDIBLE and choice.

S. cavipes is one of the more beautiful of the boletes of the West, and it is found growing in association with larches. Its unique specific character is the hollow stipe, which sets it off from all other species of *Suillus*. The densely tomentose or scaly pileus bears a resemblance to that of *S. lakei* (see below), but that species has a solid stipe.

Suillus granulatus (Fr.) Kuntze

PILEUS up to 10 cm in diameter; convex; smooth; viscid or glutinous; pallid to flesh-color when young, becoming orange-brown with age. FLESH white to pale yellow, does not change color when cut; soft; odor and taste mild. TUBES whitish with small drops of milky or pinkish fluid exuding when young, becoming yellow and spotted with age. PORES small; somewhat angular; elongate. STIPE up to 8 cm long and from 1-2 cm in diameter; usually equal; white when young and covered with pinkish-brown glandular dots, but becoming quite yellow, especially near top, with age; does not change color when cut or bruised; annulus lacking. SPORES cinnamon; oblong; smooth; 7-10 x 2.5-3.5 μ. EDIBLE.

S. granulatus is a widespread species associated with conifer forests in many parts of the West. It may be confused with *S. pungens* (see below), which usually has some grayish tones in the pileus, as well as white tissue on the margin, and which has a strong, pungent odor.

S. albidipes (Pk.) Sing. may be confused with both *S. granulatus* and *S. pungens*, but, like *S. brevipes* (see above), glandular dots are absent or present only in very small numbers on the lower part of the stipe. *S. albidipes* differs from all these species by the presence of a cottony mass on the margin of the pileus, especially noticeable when young. It is edible.

Suillus grevillei (Kl.) Sing.   Plate 29

PILEUS 5-16 cm in diameter; broadly convex to nearly plane; surface smooth and very glutinous; bay or chestnut-red, with narrow band of bright yellow around the margin.

FLESH pale yellow when young; becomes flesh-color or even reddish when cut; odor mild. TUBES pale to bright yellow, becoming pinkish-brown where bruised. PORES small; angular. STIPE relatively short and thick, 4-10 cm long and up to 4 cm in diameter; equal and solid; yellow at first, with extension of the tubes appearing as lines from apex to annulus, soon becoming blotched with reddish-brown below; annulus superior and somewhat gelatinous. SPORES olive-brown to dull cinnamon; oblong; smooth; 8-10 x 3-3.5 μ. EDIBLE.

S. grevillei, like *S. cavipes* (see below), is restricted to the more northern areas where larches grow. It is a striking-looking bolete with its bright, glistening, chestnut-colored pileus usually rimmed with bright yellow. Although edible, it is not regarded as choice by those who have tried it.

Suillus lakei (Murr.) Smith and Thiers

PILEUS up to 13 cm in diameter; broadly convex to plane; surface covered with reddish-brown scales and dry, but viscid beneath this layer; remains of veil usually adhering to margin. FLESH yellow, does not change when cut; odor and taste mild. TUBES ochraceous or dull orange-yellow, turning brown where bruised; pores large, angular; radially arranged and decurrent on stipe. STIPE up to 10 cm long and 3 cm in diameter; solid; equal or somewhat enlarged toward base; yellow when young, but soon stains reddish-brown, below thin, membranous annulus, which is not far below apex; flesh near base sometimes becomes bluish-green when cut; mycelium white. SPORES olive-yellow to dull cinnamon; subellipsoid to subcylindric; smooth; 7-10 x 3.5 μ. EDIBLE.

S. lakei is associated principally with Douglas Fir, although it does occur occasionally in mixed woods with other conifers, oaks, and Madrone. It somewhat resembles *S. cavipes* (see above) in the reddish, scaly surface of the pileus, but *S. lakei* has a solid stipe, and differs also in habitat.

A very similar species, **S. pictus** (Pk.) Smith and Thiers, occurs only under white pines. It has red fibrils on a yellow background. *S. pictus* is edible and choice.

Suillus pungens Thiers and Smith   Plate 30

PILEUS up to 13 cm in diameter; convex when young but becoming nearly plane with age; viscid; olive-green or gray-green when young, becoming orange-yellow, often blotched with olive or gray, with age. FLESH white when young, becoming yellow with age; does not change when cut; odor pungent, and taste disagreeable. TUBES whitish to cream-color, exuding droplets of white milk when young, becoming yellow with age, and droplets turning brown. PORES small; irregular in shape. STIPE up to 12 cm long but usually shorter, about 1-2 cm in diameter; equal; solid; white with numerous reddish-brown glandular dots; flesh does not change from white when cut. SPORES yellowish-brown; elliptic; smooth; 9.5-10 x 3-3.5 μ. EDIBLE, but undesirable; see below.

S. pungens is common wherever Monterey Pines grow or have been introduced. It occurs commonly along the southern and central California coast and may be very abundant where the pines are growing in sandy soil. *S. pungens*, like *S. granulatus* (see above), has glandular dots which readily blacken the fingers a few minutes after the stipe has been handled, and the stain is difficult to remove. The strong odor and the grayish tones to the pileus are important field characters. Its disagreeable odor and taste preclude its use as food.

Suillus sibiricus (Sing.) Sing.

PILEUS up to 10 cm in diameter; broadly convex, often becoming plane, and usually with a small umbo; remains of veil often adhering to margin; viscid; surface dull yellow with reddish-brown patches, especially noticeable toward margin. FLESH pale yellow; odor mild. TUBES yellow, becoming reddish-brown where bruised. PORES large; angular. STIPE 5-10 cm long and up to 1.5 cm in diameter, relatively slender; dull yellow; glandular dotted; annulus usually absent. SPORES pale brown; elliptic; smooth; 8-11 x 4 μ. EDIBLE.

S. sibiricus is reportedly associated only with Western White Pine in the West. It is recognizable by its rather broad, dingy-yellow pileus with brownish or reddish-brown patches on the surface; this is one of the easiest ways to distinguish it from *S. umbonatus* (see below), which is also dull yellow and usually umbonate.

Suillus subolivaceus Smith and Thiers   Plate 31

PILEUS up to 13 cm in diameter; broadly convex to plane; narrow, sterile margin extending beyond tubes; surface very glutinous; olive-gray to yellow-brown. FLESH pale gray to yellowish; does not change color when cut; odor mild. TUBES grayish-yellow, exuding numerous droplets of pale yellow liquid. PORES fairly large. STIPE relatively slender, up to 16 cm long and 2 cm in diameter; equal; whitish, becoming pale yellow near apex; glandular-dotted; annulus large, glutinous, olive-gray sheathing stipe. SPORES dull cinnamon; almost spindle-shaped; smooth; 8-11 x 3-4.5 μ. EDIBLE.

S. subolivaceus occurs in autumn in mixed coniferous forests in Oregon, Washington, and northern Idaho, especially those in which Western White Pine is found. Important field characters to look for are the glutinous pileus, the dull yellowish to grayish-yellow tubes with abundant droplets of pale yellow exudate, and the glandular dotted stipe with a large, collar-like, glutinous annulus. Another distinctive feature is the abundance of coarse, white mycelium associated with the base of the stipe.

S. luteus (Fr.) S.F. Gray, which occurs under conifers in the fall, also has a large, extremely viscid annulus that is purplish on the underside. It is dark yellowish-brown or reddish-brown, viscid, and edible.

Suillus tomentosus (Kauf.) Sing., Snell, and Dick
Plate 32

PILEUS up to 16 cm in diameter; convex, becoming more expanded with age; surface covered with coarsely tomentose to scaly, grayish to brown layer when young, becoming rather smooth with age; viscid surface beneath tomentose layer, ranging from pale yellow to orange-yellow. FLESH thick; pale yellow to pale orange-yellow, generally changes to blue when cut; odor and taste mild. TUBES vinaceous-brown to yellow-brown on surface; often stain slightly blue when cut. PORES very small; not noticeably angular. STIPE 5-10 cm long, up to 3 cm thick; tends to become enlarged basally when long; pallid to orange, sometimes stains blue when cut; numerous grayish to brown glandular dots, especially near apex; annulus lacking; mycelium white. SPORES dark olive; fusiform; smooth; 7-12 x 3-5 μ. EDIBLE and good.

S. tomentosus is widespread in western United States and southern Canada but seems to be most often associated with Lodgepole Pine in the mountains and Beach Pine along the coast. One of its principal distinguishing features is the pale vinaceous-brown color of the tubes when young. This, combined with the tomentose surface of the cap, the absence of an annulus, and the presence of gray to brown glandular dots on the stipe, serve for ready identification. It fruits from midsummer to late autumn, depending on the locality.

Another form, *S. tomentosus* var. **discolor**, has been named from the Rocky Mountains, Its pileus is a duller greenish-yellow, becoming olive-brown to cinnamon-brown with age.

Suillus umbonatus Dick and Snell

PILEUS up to 8 cm in diameter; broadly convex to plane, usually with an umbo; surface uneven; viscid; pale yellow to olive-buff. FLESH pale yellow, does not change color when cut; soft; odor mild. TUBES pale greenish-yellow, staining brownish where bruised. PORES large; irregularly angular. STIPE relatively small, 4-8 cm long and 4-8 mm in diameter; whitish with yellowish glandular dots; gelatinous annulus. SPORES dull cinnamon; elliptic; smooth; 7-10 x 4-4.5 μ. EDIBLE.

S. umbonatus occurs in summer and autumn in the western mountains and in October and November along the coast. It is most likely to be confused with *S. sibiricus* (see above), and the two species may be found in the same general area where Lodgepole Pine and Western White Pine occur. However, *S. umbonatus* is an associate of the Lodgepole and Beach Pine and not the Western White Pine like *S. sibiricus*, and further differs by having a gelatinous annulus and lacking reddish patches on the pileus.

Genus *Fuscoboletinus*

Fuscoboletinus is closely related to *Suillus*, but differs from it by radially arranged pores and by spores that vary from purplish-brown or chocolate-gray to reddish. Only two species are likely to be encountered in the West by the nonprofessional collector. Both are associated with Western Larch (*Larix occidentalis*).

Fuscoboletinus ochraceoroseus (Snell) Pom. and Smith

PILEUS usually large, 8-15 (25) cm in diameter; broadly convex; sterile membranous margin; surface dry and scaly; varies from yellow with pinkish or rosy tinge to bright red.

FLESH thick; pale yellow except just beneath surface, where it become pink; odor and taste rather acrid. TUBES yellow, usually with pinkish tinge when young, gradually turning vinaceous-brown and ultimately deep chocolate with age. PORES large; angular; radially arranged; decurrent. STIPE relatively short, 3-5 cm in length, slightly more than 3 cm thick; usually enlarged at base into small bulb; tubes decurrent on apex to or almost to annulus; upper part similar in color to tubes; white to yellow or orange-brown below annulus; annulus membranous. SPORES dark vinaceous-brown; smooth; elongate, almost cylindrical; 7.5-9.5 x 2.5-3 μ. EDIBLE, but poor.

F. ochraceoroseus is one of the most beautiful boletes in the West, but it is restricted to larch forests in the Northwest, where it is found in summer and autumn. The color of the dry, scaly pileus is quite variable: in rather open, sunny situations it may be almost straw-yellow, with only a tinge of pink, but in damp situations under dense conifers it may be a brilliant brick-red. It is reported to be bitter when cooked.

Fuscoboletinus aeruginascens (Secr.) Pom. and Smith

PILEUS 3-12 cm in diameter; convex at first but later expanding; surface very glutinous; narrow, sterile membranous margin; gray, often with vinaceous tinge, sometimes becoming greenish or yellowish in age. FLESH white to pale yellow, turns bluish-green when cut; odor and taste mild. TUBES white to grayish, becoming more brownish-gray with age and bruising bluish. PORES large; angular; radially arranged. STIPE up to 8 cm in length and 2 cm in diameter; often tapering upward; viscid; surface reticulate above high annulus; pallid with greenish tinge above annulus, grayish-brown below. SPORES vinaceous-brown; smooth; elliptical; 8.5-14 x 3.5-5 μ. EDIBLE and good.

F. aeruginascens, also restricted to larch forests in the Northwest, is very different in appearance from the beautiful

F. ochraceoroseus (see above), but the radially arranged tubes and vinaceous-brown spores indicate relationship.

Genus *Tylopilus*

In *Tylopilus* the pileus may be smooth or rough, dry or viscid. The tubes are whitish to pale gray when young, becoming flesh-colored to dark brown or almost black with age. The pores are angular and variable in size. The stipe lacks glandular dots, although it may be reticulated or somewhat rough. *Tylopilus* differs from *Leccinum* (see above) in the absence of conspicuous ornamentation on the stipe and by having pinkish or vinaceous-brown to chocolate rather than yellowish to olive-brown spores. Species previously assigned to the genus *Porphyrellus* are now included in *Tylopilus*. Several species occur in the West, but only *T. pseudoscaber* is likely to be encountered by the collector.

Tylopilus pseudoscaber (Secr.) Smith and Thiers

PILEUS 5-15 cm in diameter; convex to nearly plane; surface dry or somewhat moist, but not viscid; somewhat tomentose; varying from olive-gray to brownish-black; surface often cracked. FLESH white, becoming dark with age; turns bluish when cut; odor and taste slightly acrid. TUBES white or grayish, turning blue when cut. PORES rather large. STIPE 5-15 cm in length, up to 4 cm in diameter; finely ridged longitudinally; brownish-black or sooty. SPORES reddish-brown; quite elongate, almost fusiform at times; smooth or minutely spiny; 14-18 x 6-7 μ. EDIBLE.

T. pseudoscaber has widespread distribution in coniferous forests but is most abundant in the Pacific Northwest. Because of its dark color, it is often referred to as the Black Bolete. Despite the somewhat medicinal or acrid odor of the flesh, it is reported to be edible.

Genus *Pulveroboletus*

Pulveroboletus species are distinctive, but rarely encountered in the West. They are terrestrial and, when young, are clothed with a conspicuous yellow, powdery material which is thought to be a remnant of the universal veil. The pileus is viscid when wet, and the stipe lacks glandular dots.

Pulveroboletus flaviporus (Earle) Sing.

PILEUS 7-12 cm in diameter; broadly convex; viscid; surface faintly reticulate, with reddish-brown ridges on vinaceous or pinkish background. FLESH white and unchanging; odor and taste mild. TUBES bright, golden-yellow, bruising dusky-gray or olive. PORES medium-sized. STIPE similar in color to pileus but becoming whitish above, and then yellow at apex, because of tubes extending down as thin lines; base tapers downward as rootlike extension with white mycelium; annulus absent. SPORES brown; ovoid to ellipsoid; smooth; 13-18 x 5-7 μ. EDIBILITY unknown.

P. flaviporus has been found in oak woodlands in central California from Santa Clara County north at least to Napa County.

Genus *Boletus*

Boletus is a well-known genus, because many of the most popular choice edible boletes are found here. It is characterized by mostly medium to large mushrooms that lack a veil; have a smooth, scaly, or reticulate stipe; and cinnamon, yellow-brown, olive-brown, or dark brown spores. Species of *Boletus* are almost invariably terrestrial; two exceptions are **B. mirabilis**, which grows on decaying wood, and **B. parasiticus**, which parasitizes species of another fungus, *Scleroderma*. We have no authenticated reports of the occurrence of *B. parasiticus* in the West, although its host is found here.

The world-famous *Cèpe*, *B. edulis*, is one of the few wild mushrooms which are allowed to be sold, either dried or canned, in North America. However, there are some poisonous boletes, and those with red tube mouths are particularly suspect.

KEY TO SPECIES OF *BOLETUS*

1a. Growing on the ground........................2
 b. Growing on or near
 decaying wood................. *B. mirabilis,* p. 108
   2a. Pores red, orange-red, or reddish-brown 3
    b. Pores at maturity yellow, olivaceous, or
     reddish-brown (sometimes white
     when young) 6
3a. Stipe reticulate................................4
 b. Stipe not reticulate 5
   4a. Stipe clavate; pileus brown or
    reddish-brown...........*B. pulcherrimus,* p. 103
    b. Stipe bulbous and constricted at apex;
     pileus pale olive or
     olive-buff .... *B. satanus,* p. 102
5a. Pileus reddish or reddish-brown; fresh
 flesh stains blue.............. *B. erythropus,* p. 104
 b. Pileus clay-color to cinnamon;
 flesh does not stain blue; yellow
 mycelium at base .... *B. piperatus,* p. 103
   6a. Stipe reticulate 7
    b. Stipe not reticulate 8
7a. Pileus pallid to reddish-brown;
 stipe massive .... *B. edulis,* p. 105
 b. Pileus rosy pink .... *B. regius,* p. 105
   c. Pileus olivaceous
    to yellow-brown .... *B. subtomentosus,* p. 106

8a. Stipe yellow to orange-yellow,
pileus olive-brown to
dark brown .... *B. calopus*, p. 107
  b. Stipe partly red or brownish **9**
9a. Pileus dark brown to
blackish-brown .... *B. zelleri*, p. 107
  b. Pileus olive-buff with pinkish or
reddish overtones .... *B. smithii*, p. 104
  c. Pileus dark olive to olive-brown; areolate,
with red cuticle usually apparent
between cracks .... *B. chrysenteron*, p. 106

Boletus satanus Lenz

PILEUS up to 20 cm in diameter; convex when young,
expanding with age; surface usually smooth; dry; margin
often irregular; pale olive or olive-buff, sometimes pinkish
near margin. FLESH thick; whitish to pale yellow; turns blue
when cut, when young. TUBES greenish-yellow; turns blue
when cut. PORES red; very small; slightly angular. STIPE up
to 16 cm long, 8 cm or more in diameter; abruptly narrowed
at apex, bulbous below; surface dry; reticulate; concolorous
with pileus. SPORES olive-brown; elongate; smooth; 11-15 x
3.5-7 μ. POISONOUS.

B. satanus has been recorded only in California, where it
occurs under oaks in autumn and early winter. It is readily
recognized by its large size, red tube mouths, and abruptly
bulbous stipe. The latter character separates it readily from
B. pulcherrimus (see below). Both are poisonous, although
some people say *B. satanus* is edible if it is first parboiled
and the water discarded.

Boletus pulcherrimus Thiers and Halling   Plate 33
Boletus eastwoodiae (Murr.) Sacc. and Trott.

PILEUS 10-16 cm in diameter; convex when young, becoming plane with age; smooth to tomentose; dry; buffy-brown with reddish tinge, especially near margin. FLESH thick; yellow; turns blue when cut; odor and taste mild. TUBES yellow; become blue when cut. PORES deep red; small; angular. STIPE up to 16 cm long, 5 cm in diameter; equal or enlarged toward base; surface dry; with red reticulations on orange-buff base. SPORES olive-brown; elongate; smooth; 13-16 x 4.5-6.5 μ. POISONOUS.

B. pulcherrimus is rare, and it occurs only along the Pacific Coast in mixed hardwood-conifer forests. *B. satanus* is similar, but it grows only under oaks.

Boletus piperatus Fr.

PILEUS up to 6 cm in diameter; convex when young, becoming plane with age; surface smooth; viscid when wet, but otherwise dry and smooth; yellowish-brown to reddish-brown, with color very uniform over surface. FLESH brownish-yellow, unchanging; taste, very acrid. TUBES yellowish-brown. PORES reddish-brown; large; angular. STIPE up to 8 cm in length; slender; smooth; reddish-brown above, becoming yellow at base; mycelium yellow. SPORES yellowish; elongate; smooth; 10-14 x 3-4 μ. NOT POISONOUS, but acrid.

B. piperatus, though not abundant in the West, is widely distributed. It is most frequently found where there are spruces. Despite its acrid taste, it is not considered poisonous, and it may be cooked in small quantities with other species.

Boletus erythropus (Fr.) Krombh.

PILEUS up to 16 cm in diameter; convex when young, becoming plane with age; surface dry; tomentose; dark brown. FLESH yellow; becomes blue when cut; odor and taste mild. TUBES yellowish-green, becoming blue when cut. PORES small; yellow when young, becoming red with age. STIPE up to 16 cm long, 3 cm in diameter; somewhat enlarged toward base; reddish; mycelium white. SPORES olive-yellow; elongate; smooth; 13-18 x 5-6.5 μ. POISONOUS.

B. erythropus occurs in deciduous, coniferous, and mixed forests throughout the temperate zones of the world. Although eaten by some, it can be mildly poisonous and should be avoided.

Boletus smithii Thiers

PILEUS up to 16 cm in diameter; broadly convex when mature; surface dry; fibrillose to almost tomentose; pale olive-buff, with pinkish tinge or patches of pink; pink or reddish tones becoming more evident with age. FLESH pale yellowish-green; sometimes changes to blue when cut; odor and taste mild. TUBES yellow, turning blue when cut. PORES medium size; angular; yellow. STIPE up to 16 cm in length, 3 cm in diameter; usually enlarged slightly toward base; dry; smooth; reddish above, becoming yellow or whitish below, or at times the colors are reversed. SPORES olive-yellow; elongate; smooth; 14.5-19 x 4-6 μ. EDIBILITY unknown.

B. smithii has been found in coniferous and mixed forests in Idaho, Washington, Oregon, and northern California. This beautiful species was described by Dr. Harry Thiers and named for Dr. Alexander H. Smith, who has contributed so much to our knowledge of western fungi.

A closely related species, **B. flavifolia**, differs in having yellow reticulations on the stipe.

Boletus regius Krombh. Plate 34

PILEUS up to 20 cm in diameter; convex when young, expanding irregularly with age; surface usually dry, with numerous depressions; somewhat tomentose when young; pink to reddish. FLESH bright yellow; irregularly changes to blue when cut; odor and taste mild. TUBES bright yellow, usually turning blue when cut. PORES medium size; angular; similar in color to tubes. STIPE up to 13 cm in length, 5 cm thick; somewhat enlarged toward base; surface reticulate; yellow. SPORES olive-brown; elongate; smooth; 12-17 x 4-5 μ. EDIBLE and choice.

B. regius occurs in both coniferous and mixed forests in California. This large, colorful bolete is considered almost as choice eating as *B. edulis* (see below). Chief field characters are its large size, rosy pileus, and bright yellow tubes.

Boletus edulis Bull. ex Fr.

PILEUS up to 20 cm in diameter; convex, expanding with age; moist to almost viscid; surface often wrinkled or uneven; color variable, pallid to tawny-brown, sometimes with vinaceous tinge. FLESH white, thick; odor and taste pleasant. TUBES white when young, becoming greenish-yellow with age. PORES small; similar in color to tubes. STIPE 10-18 cm long, up to 7 cm thick; enlarged toward base; surface with white or brown reticulations; pallid to brownish, paler near pileus. SPORES olive-brown; elongate; smooth; 13-17 x 3.5-6 μ. EDIBLE and choice; see below.

B. edulis fruits in autumn in the West, from Alaska south, where it is associated with a wide variety of coniferous forests. In France it is known as *Cèpe*, and in Germany it is called *Steinpilz*; it is regarded as one of the most delicious of all mushrooms. Many people dry these boletes after cutting them up into medium-thin slices; they preserve well and are excellent for flavoring.

Boletus chrysenteron Fr.

PILEUS up to 8 or more cm in diameter; convex to broadly convex; surface dry, velvety; cracks or splits with age to expose lighter flesh beneath, which may become pinkish-red near margin; pale to dark brown. FLESH white to pale yellow, except near surface at margin, where usually tinged red; sometimes changes to blue when cut. TUBES yellow, usually turning blue where bruised. PORES large; irregular in shape. STIPE up to 10 cm long, 2 cm in diameter; equal; smooth, but sometimes with minute longitudinal ridges; yellow tinged with red, especially middle and lower part. SPORES olive-brown; elliptic; smooth; 12-14 x 5-6 μ. EDIBILITY questionable.

B. chrysenteron is a widely distributed species found in coniferous forests, oak or oak-madrone woodlands, and mixed coniferous and hardwood forests. It is closely related to *B. zelleri* (see below), which is more abundant in the West. *B. chrysenteron*, however, has a cracked pileus of a much lighter color, and the stipe is generally less red.

Another very similar species, **B. truncatus** (Sing., Snell, and Dick) Pouzar, also occurs on the Pacific Coast and can be distinguished with certainty from *B. chrysenteron* and *B. zelleri* by the shape of the spores, which are flat or truncate at one end. Its edibility is unknown.

Boletus subtomentosus Fr.

PILEUS up to 8 cm in diameter; broadly convex to plane, or even centrally depressed with age; surface minutely tomentose; dry; olive to light brown; cracks progressively from margin to center, exposing flesh beneath. FLESH creamy-white to pale yellow; sometimes turns blue when cut; firm; odor and taste mild. TUBES bright yellow when young, becoming brownish-yellow with age, bruising blue at times. PORES large; angular; decurrent on stipe as ridges which become reticulate. STIPE up to 10 cm long, 2 cm in diameter; solid; more or less equal; yellowish, brightest near pileus;

surface slightly reticulate; flesh white, does not change color when cut; mycelium at base usually yellow. SPORES olive-brown; fusiform to elliptic; smooth; 12-16 x 3.5-5 μ. EDIBILITY unknown.

B. subtomentosus, while not particularly abundant, is found in conifer forests along the Pacific Coast and in the Rocky Mountains.

Boletus zelleri Murr.   Plate 35

PILEUS up to 10 cm in diameter; convex when young, sometimes becoming plane with age; surface dry, velvety; dark brown to almost black; margin frequently reddish. FLESH pale yellow to white; sometimes turns blue when cut; odor and taste mild. TUBES usually bright yellow, sometimes staining blue when bruised; somewhat depressed next to stipe, extending down latter as lines. PORES moderately small, subangular. STIPE up to 10 cm long, 2 cm in diameter; equal or tapering slightly toward upper end; surface dry, relatively smooth although it may appear finely ridged longitudinally and covered with minute granules; yellow above, but red for most of length at maturity. SPORES yellow to tawny; subellipsoid; smooth; 12-15 x 4-5.5 μ. EDIBILITY unknown.

B. zelleri is common in Coast Redwood forests along the Pacific Coast and in mixed coniferous forests in the Sierra Nevada, Cascades, and Rocky Mountains. Fruiting bodies may be solitary or occur in small groups.

B. zelleri is similar to *B. chrysenteron* (see above), but the surface of the pileus is usually darker, more velvety, and rarely cracked. Spore shape readily distinguishes it from *B. truncatus* (see above).

Boletus calopus Fr.

PILEUS 10-26 cm in diameter; occasionally even larger; convex with incurved margin; dry; fibrillose; olive-brown to

dark yellow-brown in age. FLESH yellow or whitish; quickly
turns bright blue when cut; taste bitter. TUBES pale yellow to
greenish-yellow. PORES small; yellow; bluing. STIPE 10-15
cm long; equal or bulbous; yellow to orange-yellow, pinkish at base; finely reticulate. SPORES olive-brown; sub-
ventricose; 13-19 x 5-6 μ. POISONOUS.

B. calopus is solitary to gregarious in conifer forests,
especially in the higher mountains in summer. It is one of
the large boletes, but unfortunately it is not edible.

Boletus mirabilis Murr.

PILEUS 7-15 cm in diameter; convex when young, becoming nearly plane with age; fibrillose to tomentose, becoming
conspicuously scale at maturity; dark reddish-brown. FLESH
white to yellow; taste mild to slightly acrid; odor not
distinctive. TUBES yellow, becoming olive in age. PORES
round to angular and same color as tubes. STIPE 8-15 cm
long, 1-3 cm thick at apex and clavate, up to 5 cm thick at
base; reticulate at apex and usually roughened or pitted
toward base; dark brown. SPORES olive-brown, ellipsoid,
smooth, 18-22 x 7-9 μ. EDIBLE and choice.

B. mirabilis occurs in the coastal forests as well as the
northern Rocky Mountains on soil or rotten wood, and is
usually associated with species of hemlock. It is common in
Oregon and Washington.

PUFFBALLS AND THEIR ALLIES/Gasteromycetes

Puffballs, earthstars, bird's-nest fungi, stinkhorns, and a few other small groups are known as the Gasteromycetes, or stomach fungi. The fruiting bodies in this diverse category are more or less globular, at least at first. The spores are produced on basidia, usually along with strands of sterile hyphae, the *capillitium*, and this mass is called *gleba*. It is contained within the *peridium*, an outer covering which may be single or multilayered. At maturity the gleba usually becomes dry and powdery and the peridium ruptures in various ways, thereby releasing the spores for dispersal in the wind, as they are not forcibly discharged.

An exception is the family Phallaceae, commonly called stinkhorns, in which the gleba is a slimy, foul-smelling mass. An egglike peridium is ruptured by the growth of an elongate stalk, the *receptaculum*, or receptacle; the odor of the gleba then attracts insects which eat it and disseminate the spores on their legs or in their droppings.

Some Gasteromycetes possess a stipelike sterile base or a distinct stipe (stalked puffballs).

FAMILY GEASTRACEAE (Earth Stars)

The Geastraceae are puffball-like fungi with a peridium composed of several layers. The outer two or three layers, the *exoperidium*, separate from the *endoperidium* and rupture in a stellate manner. At maturity the endoperidium either disintegrates or opens by one or more pores, releasing the spores, which are ornamented.

Genus *Geastrum*

The fruiting body in *Geastrum* is globose or nearly so, with or without a short stalk. The exoperidium ruptures into star-shaped rays surrounding the inner spore sac, which ruptures with a single apical pore to release the spores at maturity.

Geastrum coronatum Pers.

FRUITING BODY up to 2 cm in diameter; globose at first; exoperidium ruptures into 4 to 6 pointed rays, which recurve to raise inner spore sac above ground; tips of rays remain attached to mycelial cup beneath. SPORE SAC somewhat oval; raised on a short pedicel; surface rough; dark brown; apical pore marked by a distinct groove. SPORES brown; globose; ornamented; 3.5-6 μ. EDIBILITY unknown.

   G. coronatum (Fig. 13) is one of the common species of

FIG. 13. *Geastrum coronatum.*

small earthstars found in summer and fall in coniferous forests of the Rocky Mountains and Pacific Coast. It is easily overlooked because of its brown coloration, which is similar to the coniferous duff in which it grows.

Geastrum triplex Jung.

FRUITING BODY up to 5 cm across; globose at first; exoperidium splits into 4 to 8 rather thick, fleshy rays that recurve; base of rays is depressed, forming a cup in which spore sac sits. SPORE SAC globose; tan; surface smooth; apical pore surrounded by radially arranged fibrils. SPORES brown; globose; ornamented; 3.5-4.5 μ. EDIBILITY unknown.

G. triplex is most often found along the Pacific Coast in nonconiferous woodlands. The fleshy rays; absence of a pedicel, or stalk, under the spore sac; and the cup in which the sac is contained are distinctive characters.

FAMILY LYCOPERDACEAE (Puffballs)

The fruiting body in the Lycoperdaceae is round to pear-shaped. The exoperidium either disintegrates or breaks away in sections; the endoperidium may rupture by an apical pore or break open irregularly. At maturity the glebal mass is dry and powdery. Spores may be smooth or ornamented.

Puffballs may be terrestrial or lignicolous. Some common species, representing three genera found in the West, are given here.

Genus *Calbovista*

Calbovista is a terrestrial genus. The fruiting body is medium to large; the outer peridium breaks into thick, often pyramidal plates which fall off at maturity, and the gleba is dark olive-brown at maturity. Spores are brown and globose. The capillitium with thornlike or antler-like branches distinguishes members of this genus from those of *Calvatia* (see below).

Calbovista subsculpta Morse

FRUITING BODY up to 15 cm in diameter; round to oval; exoperidium white when young, covered with low wartlike scales or pyramids; scales break off at maturity, exposing the olive-brown endoperidium; gleba white when young, becoming dark brown at maturity; capillitium has thornlike branches; endoperidium disintegrates at maturity. SPORES round; yellowish-brown; smooth; 3-5 μ. EDIBLE.

C. subsculpta is a common species in late spring and summer in the higher coniferous forests of the West, occurring from Alaska and the Yukon south to California. It is often found around mountain cabins where the soil becomes packed. The Sculptured Puffball, *Calvatia sculpta* (see below), which has long, pointed, pyramid-shaped warts and does not have thornlike branches on the capillitium, may grow in the same areas.

Genus *Calvatia*

Calvatia contains medium to very large puffballs, varying from round to pyriform. The exoperidium may be smooth, rough, warted, or with pointed pyramidal scales; the endoperidium is thin and fragments at maturity after the exoperidium scales off. The base of the fruiting body often consists of sterile tissue which may remain intact long after the spores are discharged; the capillitial threads are either unbranched or sparsely branched.

Calvatia cyathiformis (Bosc) Morg.

FRUITING BODY up to 20 cm high; generally pear-shaped; exoperidium smooth when young, but cracks into small plates which drop off at maturity; endoperidium very thin, fragmenting at maturity; gleba white when young, becoming deep purple with age; sterile base remains as a cuplike structure after maturity. SPORES purple; globose; minutely ornamented with spines; 3.5-7.5 μ. EDIBLE when young.

C. *cyathiformis* can be found in open grassland and prairies in spring and early summer in the West. The large, cup-shaped sterile base, with some of the purple gleba still adhering, is more often found than the young fruiting bodies. When young and white throughout, it is edible.

Calvatia sculpta (Hark.) Lloyd   Plate 36

FRUITING BODY up to 15 cm in height; varying in shape from oval to pear-shaped; exoperidium white, consisting of elongate pyramid-shaped warts terminating in a spinelike tip; warts usually with fine parallel lines on each side; endoperidium very thin, fragmenting as the outer pyramidal scales fall off; gleba white when young, becoming olive-brown at maturity; medium to elongate sterile base; attached to ground by white mycelial threads. SPORES olive-brown; round; minutely ornamented; 3.5-6.5 μ. EDIBLE when young.

C. *sculpta* is one of the most beautiful of the western puffballs. When fresh it resembles a globular mass of meringue, often on a basal pedicel, from which many pyramid-shaped points protrude. Most of these pyramids show fine parallel lines of growth on their sides. These sculptured puffballs are found in the litter of coniferous forests, especially where there is granitic sand, at middle to high elevations in late spring and summer in the Sierra Nevada as well as the Cascades. In the Rocky Mountains they have even been recorded in early fall. They are associated with the morel season.

C. *sculpta* might be confused with *Calbovista subsculpta* (see above), which often grows in the same localities. While both have pyramidal warts, those of C. *sculpta* are attenuated into long spines; and the capillitial threads of *Calbo-vista subsculpta* show conspicuous antler-like processes lacking in *C. sculpta*—to determine this, one must have mature specimens. *C. sculpta*, like most puffballs, is excellent for eating when young and the gleba is white and firm.

Calvatia subcretacea Zeller

FRUITING BODY up to about 5 cm in diameter; round to oval in shape; exoperidium thick, covered with small white warts with smoky gray tips that break up at maturity; endoperidium thin, fragmenting; gleba white when young, becoming dark olive-brown at maturity. SPORES olive-brown; round; smooth; 3.5-6.5 μ. EDIBLE when young.

C. subcretacea is another high mountain species of the coniferous forests of the West. This puffball grows in the same general areas and at the same seasons of the years as *C. sculpta* and *Calbovista subsculpta*. Its golf-ball size and thick peridium covered with small white warts that are gray at their tips are good characters for identification.

Genus *Lycoperdon*

Lycoperdon contains the common small to medium-sized puffballs that are round to pear-shaped. The exoperidium is smooth or covered with small spines or warts that flake off at maturity; the endoperidium is membranous and maintains its shape after maturity. Spores are released through an apical pore. The gleba is white when young, becoming olive-brown when mature. A sterile base is usually present, and the fruiting body is attached to the substratum by means of one or more *rhizomorphs*, or stringlike masses of hyphae.

Members of this genus are the commonest puffballs in the West. During the cool moist seasons they are found on lawns, in pastures, in uncultivated grasslands, broadleafed woods, and coniferous forests. Most are terrestrial, but some grow on rotting wood. Three very common species are described here.

Lycoperdon perlatum Pers.
Lycoperdon gemmatum Batsch

FRUITING BODY up to 7 cm high; pear-shaped; exoperidium composed of numerous small, white to pale tan, cone-shaped spines which fall off at maturity; endoperidium smooth, but often showing spots where spines were formerly attached; gleba white, becoming olive-brown with age; well-developed sterile base with large chambers present. SPORES olive-brown; round; with minute spines; 3.5-4.5 μ. EDIBLE when young; see below.

L. perlatum is a very common species in both broad-leafed and coniferous woods in spring, late summer, and autumn, depending upon the weather. It even occurs on the Alaskan tundra. It may be solitary, but more often grows in clumps in humus. The spores emerge slowly from the apical pore after maturity, and some may remain within the endoperidium for months.

FIG. 14. *Lycoperdon perlatum.*

L. perlatum (Fig. 14) is excellent for eating if collected when it is young and the glebal mass has not begun to turn yellow. Cooked with milk and the proper seasoning, it makes an excellent chowder.

Lycoperdon pyriforme Pers.

FRUITING BODY up to 3 cm high; pear-shaped; variable in color, but usually light tan when young, becoming darker with age; sterile base whitish; exoperidium composed of minute spines or scales, often difficult to see, and often adhering to endoperidium until maturity; endoperidium smooth; gleba white when young, becoming olive-brown with age; sterile base often large, composed of small chambers. SPORES olive-brown; round; smooth; 3-3.5 μ. EDIBLE when young.

L. pyriforme is a common, very widespread puffball. It grows on rotting wood in both broadleafed and coniferous forests in summer and fall in the Rocky Mountains, and in fall along the Pacific Coast. It differs from *L. perlatum* (see above) in its smoother surface, tan color, smaller size, and lignicolous habit. Fruiting bodies often grow in dense clusters. Like most puffballs, it is edible when it is young and the gleba is white.

Lycoperdon pusillum Pers.

FRUITING BODY 1 to 2 cm in diameter; round; peridium smooth, white; gleba white when young, becoming brown with age; sterile base lacking. SPORES brown; round; with minute spines; often with a pedicel attached; 3.5-4.5 μ. EDIBILITY unknown.

L. pusillum is the smallest of our common puffballs. It is generally found in moist and cool grassy areas in summer and fall in the western states and provinces. Its small size, spherical shape, smooth white surface, and absence of a sterile base are good field characters.

FAMILY NIDULARIACEAE (Bird's-nest Fungi)

The common name of bird's-nest fungi was bestowed upon the family Nidulariaceae because the cup- or urn-shaped fruiting bodies have tiny oval *peridioles* which look like eggs in a miniature nest. When young, there is a lid called the *epiphragm* covering the "eggs," but at maturity it ruptures and makes it possible for the spore-bearing peridioles to be seen. They may be attached to the inner surface by a thin cord called the *funiculus*, or they may be embedded in a sticky substance. The gleba may be eaten by insects and the spores deposited in their droppings, or the eggs may be scattered by the splashing of raindrops and, as they gradually disintegrate, the spores are released. Bird's-nest fungi are widespread and occur on the ground, decaying wood, berry canes, etc., but they may often be overlooked because of their diminutive size.

KEY TO COMMON GENERA OF NIDULARIACEAE

1a. Peridioles embedded
  in mucilage...................... *Nidula*, p. 117
 b. Peridioles attached to nest by a cord 2
  2a. Peridioles whitish........... *Crucibulum*, p. 118
  b. Peridioles gray to black .... *Cyathus*, p. 119

Genus *Nidula*

The fruiting body in *Nidula* is cup-shaped to vase-shaped and composed of three layers. The peridioles, attached to the inner wall of the peridium by a sticky secretion, are lens-shaped and light brown.

Nidula candida (Pk.) White

FRUITING BODY up to 1.5 cm high; cylindrical, tapering toward base; peridium thick, outer surface whitish to pale brown, velvety. PERIDIOLES pale brown; flattened; embedded in a mucilage, but not attached to inner surface of cup, which is smooth. SPORES colorless; ellipsoid; smooth; 6-10 x 4-8 μ. EDIBILITY unknown.

N. candida is a common species along the Pacific Coast from central California to British Columbia. It occurs in the fall and early winter on dead twigs and the old canes of blackberry and Salmonberry. Usually a number of fruiting bodies will be found together and at different stages of development, some with the epiphragm intact, others open with the peridioles exposed and waiting to be splashed out with raindrops, and still others with empty cups. The cups remain intact for a long time after maturity.

Genus *Crucibulum*

In members of *Crucibulum* the fruiting body is cylindrical and short. The peridium is thick and velvety on the outer surface when young, and consists of only one layer. The peridioles, attached to the inner wall of the peridium, are lens-shaped and a pale cream-color.

Crucibulum vulgare Tul.

FRUITING BODY up to 1 cm high; cylindrical but slightly narrower toward base; outer surface somewhat velvety to smooth, yellowish to cinnamon-brown; inner surface smooth, whitish. PERIDIOLES pallid, attached to inner wall of cup by a funiculus. SPORES colorless; ellipsoid; smooth; 7-10 x 4-6 μ. EDIBILITY unknown.

C. vulgare is a widespread, common species found growing on old twigs, rotting wood, manure, and vegetable debris in the fall.

Genus *Cyathus*

The fruiting body in *Cyathus* is vase-shaped. The peridium is composed of three layers, with the outer surface ranging from smooth to velvety, tan to cinnamon-brown, becoming gray-brown in age. The peridioles are lens-shaped, thin, gray to black, and attached to the inner surface of the cup by a funiculus.

Cyathus striatus Pers.

FRUITING BODY up to 1.5 cm high; vase-shaped; outer surface of peridium tawny to brown, covered with small hairlike structures; inner surface of peridium ribbed or striate, varying from gray to blackish. PERIDIOLES grayish to black; attached to cup by cord. SPORES colorless; ellipsoid; smooth; 12-22 x 8-12 μ. EDIBILITY unknown.

FIG. 15. *Cyathus striatus.*

C. striatus (Fig. 15) grows on twigs, bit of bark, and other vegetable debris in moist, wooded situations. It is a widely distributed species of bird's-nest fungus in the West, and it is also recorded in the Yukon Territory. The vertical ribs on the inner side of the cup and the dark-colored peridioles are the best identifying characters.

FAMILY PHALLACEAE (Stinkhorns)

The Phallaceae are commonly known as stinkhorns. They derive their name from the foul-smelling, sticky glebal mass that is borne on the end of a stalk called the *receptaculum*, or receptacle. The odor serves to attract insects which disseminate the spores. The fruiting body, when young, is round to oval; it consists of a tough outer layer, which is white to pink, and an inner layer containing the receptacle. The receptable ruptures through the outer membrane, or egg covering, and grows upward as a stalk, while the membrane remains as a volva-like cup at the base. The top of the receptacle may be branched or unbranched; the glebal mass is borne on the head of the receptacle or on the branches. Some mycologists consider those stinkhorns that have a branched receptacle as a separate family, the Clathraceae.

Genus *Lysurus*

In *Lysurus* there are up to 6 tapering, armlike structures at the top of the receptacle. The glebal mass containing the spores is borne on these arms, which are shorter than the basal column and may or may not be joined together at their tips. There is a well developed volva present at the base of the receptacle. Two species of this genus have been found in southern California.

Lysurus borealis (Burt) Henn.
Anthurus borealis Burt

FRUITING BODY up to 12 cm high; stipelike stalk tapering downward to basal, white volva; white to slightly yellowish; hollow. ARMS at top of stalk; flesh-colored; usually 6; wrinkled on inner and lateral surfaces; outer surface smooth, but with longitudinal groove; about 2-3 cm long, at first touching each other terminally but later spreading. SPORES olive-green; cylindrical; smooth; produced on inner and lateral surfaces of arms; 3-4 x 1.5 μ. INEDIBLE.

L. borealis is known in California from both Santa Barbara and Los Angeles, where it grows in heavily manured ground or well-fertilized lawns and flowerbeds. It is found in the East and may be expected in parts of the Southwest.

Lysurus mokusin (Pers.) Fr.

FRUITING BODY up to 15 cm high; stalk tapering downward; hollow; angular, with 4 to 7 angles; white at base, fleshy pink above; volva white, free above, and vase-shaped. ARMS at top of stalk; as many arms as there are angles; up to 4 cm long; bright red; wrinkled on sides, usually united at tip to form lantern-like structure, but sometimes separated. SPORES olive-green; ellipsoid; smooth; produced in glebal mass in triangular spaces between adjacent arms; 3.6-4.2 x 1.5-1.8 μ. INEDIBLE.

L. mokusin has been recorded in southern California every month in the year, but is most abundant from May to October.

Genus *Phallus*

In species of *Phallus*, the receptacle is a tall, unbranched stalk that terminates in a pileus-like structure on which the foul-smelling glebal mass is borne.

Phallus impudicus Pers.   Plate 37
Ithyphallus impudicus (Berk. and Curt.) E. Fischer

FRUITING BODY up to 20 cm high; pileus-like upper portion
up to 3 cm long, surface has marked reticulum whose pits
are filled with olive-brown gleba; apex of pileus has white
circle; receptacle cylindrical, whitish, somewhat spongy;
basal volva free above, pink or white. SPORES olive; ellipsoid; smooth; 3.5-4 x 1.5-2 μ. INEDIBLE.

P. impudicus is a widespread species, but in the West we
presently know of its occurrence only in coastal California,
where it grows in midwinter in rich leafmold. It is usually
gregarious. The eggs, which are attached to the substratum
by rootlike strands, may measure as much as 5 cm in length
and are conspicuously pink when they break through the
ground. They may be collected and grown in moist humus,
but the fetid odor of the maturing fruiting bodies dictates
that they be kept out-of-doors.

FAMILY ASTRAEACEAE (False Earthstars)

The fruiting body in the Astraeaceae superficially resembles that of the earthstars (Geastraceae). The exoperidium splits stellately at maturity, and subsequently opens
when moist and closes over the spore sac when dry. The
gleba is dark brown at maturity, and the spores are round
and large, 7 μ. or more in diameter. There is but a single
genus, *Astraeus.*

Genus *Astraeus*

Although species of *Astraeus* resemble true earthstars,
they differ in several ways: the basidia are not arranged in
parallel columns; they have relatively large spores; and the
rays of the exoperidium are very hygroscopic. Some members of the genus *Geastrum* (see above) react to some extent
to moist and dry weather, but none as markedly as those of
Astraeus, in which the rays are closed over the spore case
when dry but open out flat when moistened.

Astraeus hygrometricus (Pers.) Morg.

FRUITING BODY up to 5 cm across when expanded; round at first; exoperidium thick, splitting into 7 to 15 pointed rays which are usually cracked or checkered on inner surface; spore sac gray, surface rough or hairy; irregular apical pore develops at maturity; gleba white when young, becoming dark brown at maturity. SPORES brown; round; ornamented; 7-11 μ. EDIBILITY unknown.

A. hygrometricus is a widespread species in the wet season, found in open fields, especially where the soil is sandy.

A much larger species, **A. pteridis** (Shear) Zeller, occurs in the Pacific Northwest. It may obtain a diameter of 15 cm when expanded. Its edibility is unknown.

FAMILY SCLERODERMATACEAE

The Sclerodermataceae contains a group of fungi that bear a superficial resemblance to members of the family Lycoperdaceae (see above). The fruiting body varies from round to pear-shaped and generally pushes up through the surface of the ground—but unlike the common puffballs, the peridium is quite thick and not differentiated into an outer and inner layer. Capillitial threads are either absent or rudimentary.

Genus *Scleroderma*

The fruiting body in *Scleroderma* is puffball-like and appears partly or entirely above ground at maturity. The gleba inside the tough peridium is white at first, but becomes dark at maturity. Capillitial threads are absent.

Scleroderma aurantium Pers.

FRUITING BODY up to 6 cm in diameter; round to slightly oval; peridium thick; yellowish to golden, with outer surface finely cracked; gleba white when young but turns pinkish when cut, becoming dark violet with white lines at maturity; sterile base. SPORES brown; round; ornamented; 8-13 μ. POISONOUS; see below.

S. aurantium is a common species in late summer and fall along the Pacific Coast. Many persons find these "thick-skinned" puffballs pushing up through deciduous and coniferous humus in their gardens and confuse them with members of the genus *Lycoperdon* (see above).

In Europe, this species is sometimes used sparingly as seasoning in sausage and other foods, but it contains undetermined toxins and can cause severe gastric and central-nervous-system symptoms when consumed in great quantity.

FAMILY TULOSTOMATACEAE (Stalked Puffballs)

The Tulostomataceae have a round to ovoid fruiting body at first. As it matures, the spore sac is pushed upward from the ground by a strong stalk. The exoperidium is poorly developed. The endoperidium at maturity releases the spores through an apical pore or pores, or else ruptures around the middle. Representatives of two genera found in the West are described here.

Genus *Battarraea*

The genus *Battarraea* is worldwide, but most abundant in deserts or semi-arid regions. As the fruiting body matures, the exoperidium ruptures and remains as a volva-like cup at the base of the stalk, which is long and shaggy. The endoperidium, or spore sac, is hemispheric above and concave on the underside.

Battarraea phalloides (Dicks.) Pers.

FRUITING BODY up to 20 cm high; cream-color; exoperidium ruptures peripherally and remains as a basal volva; spore sac up to 5 cm in breadth, rupturing around lower periphery to release spores; stalk tapering downward, covered with coarse scales. SPORES brown; round; ornamented with small warts; 4-7 μ. EDIBILITY unknown.

B. phalloides (Fig. 16) is most likely to be found on the deserts of the Southwest, but it is known to occur along the southern California coast and has been recorded on dry sage-covered slopes in the Yukon Territory. In the Southwest it appears most often after summer rains and persists for many months after it has dried.

FIG. 16. *Battarraea phalloides.*

B. digueti Pat. and Hariot is known from the Mojave Desert. It differs from *B. phalloides* in that the spore sac emerges from the exoperidium by an apical rather than peripheral rupture. The endoperidium also is smaller, and the spores emerge through a number of pores on the upper surface of the spore sac. EDIBILITY unknown.

Genus *Tulostoma*

Species of *Tulostoma* are small puffballs on top of a slender stalk. The exoperidium is thin and soon lost, or else becomes covered with sand. The endoperidium is paper-thin and easily broken off the stalk. The spores are released through an apical pore. A number of species have been described in North America, ranging from the southwestern deserts to the Yukon and the Subarctic.

Tulostoma simulans Lloyd

FRUITING BODY up to 4 cm high; spore sac round, up to 1.5 cm in diameter; reddish-brown when young, becoming gray with age and usually covered with sand; stalk slender, equal; mouth of spore case apical, elevated as a small tube. SPORES pale yellow; round; slightly ornamented; 4-6 μ. EDIBILITY unknown.

T. simulans is a widely distributed species occurring during the wet season in sandy soil.

A number of species of *Tulostoma* have been described from the Southwest.

T. fibrillosum White has a tawny to white spore sac and a bulblike mycelial mass which is embedded in sand at the base. It has been found growing near Desert Holly in Death Valley, California, in December. Its edibility is unknown.

4. BASIDIOMYCETES/
Club Fungi
Gilled Fungi (Agarics)

The agarics or gilled fungi comprise the majority of the fleshy fungi and contain a great many families and genera. They vary considerably in spore color, gill attachment (Fig. 17), shape, and size.

Although many choice edibles are found in this enormous group, it also contains most of the deadly poisonous species, principally in the genera *Amanita* and *Galerina*.

Since the spores of gilled fungi show a wide range in color, the following key to the families, and their arrangement in this guide, is based upon their spore color.

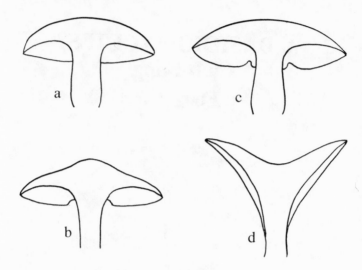

FIG. 17. Four types of gill attachment: (a) free, (b) adnate, (c) adnexed, (d) decurrent.

SPORE-COLOR KEY TO FAMILIES OF AGARICS (GILLED MUSHROOMS)

A. Spores purplish-brown to chocolate-brown
    1a. Gills free .... Agaricaceae, p. 130
      b. Gills attached .... Strophariaceae, p. 136
B. Spores pink to salmon
    1a. Spores angular or furrowed; gills
        attached .... Rhodophyllaceae, p. 145
      b. Spores smooth, elliptical; gills
        free .... Volvariaceae, p. 148
C. Spores ochre, clay-color, or brown
    1a. Spores yellowish-brown to cinnamon-brown;
        gills free .... Bolbitiaceae, p. 150
      b. Spores clay-color to brown; gills
        decurrent .... Paxillaceae, p. 152

A. SPORES PURPLISH-BROWN TO CHOCOLATE-BROWN

There are two families in this color category of spores, the Agaricaceae and the Strophariaceae. In members of the Agaricaceae, the gills are free of the stipe and always possess an annulus. The common commercial mushroom, as well as the meadow mushroom, are members of this family. Although a number of other species are edible, some may cause illness; most of the latter turn yellow on contact with a strong lye such as KOH.

The Strophariaceae contains a number of small to medium-sized species belonging to the genera *Stropharia*, *Naemataloma*, and *Psilocybe*. All have gills attached to the stipe, and many are poisonous or hallucinogenic.

FAMILY AGARICACEAE

The Agaricaceae is a small family, containing three genera, and only one, *Agaricus*, is well known. *Agaricus* is differentiated from **Melanophyllum** and **Cystoagaricus** by the filamentous hyphae in the pileus cuticle; the latter two genera have tissue composed of *pseudoparenchyma* (a form in which the hyphae have lost their individuality and are not recognizable as such).

Genus *Agaricus*

Agaricus is an easily identified genus with purplish-brown or chocolate-brown spores, free gills, an annulus but no volva, and a stipe that separates readily from the pileus. The species are a different matter, and are often difficult to differentiate. While there are several choice edible species, there is one yellow-staining group, with some poisonous species—it is believed that illness is caused by an antibiotic substance. The color change may occur spontaneously when the surface or flesh is bruised, or it may appear when lye, such as a 2.5-percent solution of KOH, is applied. Other groups may show no color reaction or produce a reddish or orange-red stain when bruised.

A. brunnescens Pk., the commercially grown mushroom that is sold in the market, was formerly called *A. bisporus* (Lange) Imbach, because it has only 2 spores on each basidium, while the wild species are 4-spored. Recently it was determined that Peck's description written in 1900 is of the same species and has priority. *A. brunnescens* and the popular Meadow Mushroom, *A. campestris*, are two of the tastiest and most commonly eaten fungi.

KEY TO SPECIES OF *AGARICUS*

1a. Flesh does not stain when bruised 2

b. Flesh becomes yellow or red when bruised 3

   2a. Growing under hardwoods or conifers in woodlands; stipe with a white floccose sheath up to the annulus .... *A. subrutilescens,* p. 132

   b. Growing in open grassland, gills at first bright pink .... *A. compestris,* p. 132

   c. Growing in lawns or often on hard-packed soil; gills at first pale pink .... *A. rodmani,* p. 132

3a. Flesh stains bright red immediately .... *A. haemorrhoidarius,* p. 133

b. Flesh stains yellow, then pinkish, and finally dull vinaceous .... *A. hondensis,* p. 133

c. Flesh stains yellow 4

   4a. Outer surface and flesh in lower stipe become bright yellow immediately when bruised .... *A. xanthodermus,* p. 134

   b. Not as above 5

5a. Pileus smooth or fibrillose 6

b. Pileus scaly 7

   6a. Fruiting body large, robust; odor of anise .... *A. arvensis,* p. 133

   b. Fruiting body medium-sized; stipe abruptly flattened or bulbous at base .... *A. silvicola,* p. 134

7a. Fruiting body very large; pileus yellow-brown with dense, darker brown scales .... *A. augustus,* p. 135

b. Pileus with minute grayish scales .... *A. meleagris,* p. 134

Agaricus campestris L. ex Fr.

PILEUS 3-8 cm in diameter; convex when young, then becoming plane; surface smooth or fibrillose; dry; white at first, becoming pale brown. FLESH thick and firm; white. GILLS free; close; bright pink at first, becoming dark brown. STIPE up to 6 cm long; equal; white; annulus single. SPORES chocolate-brown; ellipsoid; smooth; 5.5-7.5 x 3.5-5 μ. EDIBLE and choice.

A. campestris is the Meadow Mushroom, which is probably the most commonly eaten wild species in North America. It is widespread and occurs in pastures, lawns, and meadows in fall a few weeks after the first heavy rains, but may occasionally be found in spring if conditions are favorable.

A related and equally edible species, **A. rodmani**, has very firm flesh and a double ring. Its gills are a much paler pink than those of *A. campestris*, and it is most apt to be found in hard-packed soil along trails or roadsides, or even in city lawns.

Agaricus subrutilescens (Kauf.) Hotson and Stuntz
Plate 38

PILEUS up to 15 cm in diameter; convex, becoming nearly plane, often with slight umbo; surface dry, covered with vinaceous-brown fibrils, except toward margin, where underlying white flesh becomes evident. FLESH white; moderately thin; firm; lacks distinctive odor or taste. GILLS white, becoming vinaceous-pink, then brown; close. STIPE up to 15 cm long, tapering upward slightly; white to slightly vinaceous; annulus superior, thin but large; white floccose sheath extended upward to the annulus. SPORES purplish-brown; ellipsoid; smooth; 5-6 x 3-3.5 μ. EDIBLE and choice.

A. subrutilescens is fairly common in winter, especially in the Pacific coastal coniferous forests or in mixed woodlands. This beautiful agaric is often solitary.

Agaricus haemorrhoidarius Schulzer and Kalch.

PILEUS 5-10 cm in diameter; subglobose to ovoid, becoming convex to plane; usually umbonate; fibrillose to scaly; vinaceous-brown. FLESH thick; white; quickly turns red when cut. GILLS free; crowded; at first whitish, then pink, and finally purplish-brown. STIPE 15-20 cm long, 1-2 cm thick; equal or bulbous at base; whitish to brownish; staining red when bruised; annulus large, superior, persistent, white. SPORES purplish-brown; ellipsoid; smooth; 8.5-10 x 4.5-5.5 μ. EDIBLE.

A. haemorrhoidarius is found in conifer-hardwood forests of the Pacific Coast north to Alaska. The immediate reddening of the flesh when cut makes it readily identifiable. It is a good edible mushroom, though it is rare.

A. hondensis Murr. is another species of the conifer-hardwood forests of the Pacific Coast, which later develops pinkish to vinaceous stains, but this is usually preceded by a yellow color change. Its pileus is smooth or with appressed fibrils that darken to vinaceous-brown or gray with age. Unlike *A. subrutilescens* (see above), the stipe is smooth below the annulus. *A. hondensis* is poisonous, at least to some people.

The following group of yellow-staining mushrooms is complex and often so similar that they present difficulties even to experts. Some are edible, but others can cause very uncomfortable gastrointestinal illness. Therefore, be sure of correct identification before eating any of them.

Agaricus arvensis Schaeff. ex Fr.

PILEUS up to 20 cm in diameter; convex to plane in age; surface smooth; dry; creamy-white, often yellow on disc. FLESH thick; white, staining yellow; odor of anise. GILLS free; crowded; whitish at first, then grayish-pink, becoming blackish-brown. STIPE up to 20 cm long; equal or slightly bulbous at base; annulus double, with star-shaped pattern on underside. SPORES purplish-brown; elliptical; smooth; 7-11 x 4.5-5 μ. EDIBLE.

A. arvensis, the Horse Mushroom, occurs in fields and pastures in spring and fall and is conspicuous because of its size and robust stature. The anise odor is also distinctive. It is edible, but its slightly sweetish flavor is disliked by some.

Agaricus silvicola (Vitt.) Pk.

PILEUS 5-12 cm in diameter; convex, expanding to plane; surface dry; silky, fibrillose; white or with yellowish tinge in center. FLESH white, staining yellow; thick; lacks distinctive odor or taste. GILLS white, becoming pale pink, then chocolate-brown. STIPE up to 15 cm long; white, often staining yellow; annulus large, double, white, sometimes with yellow stains on underside. SPORES chocolate-brown; ellipsoid; smooth; 5-6.5 x 4-4.5 μ. POISONOUS.

A. silvicola closely resembles *A. arvensis* (see above); its main differences are it woodland habitat, somewhat smaller size, and smaller spores.

A similar mushroom, **A. xanthodermus** Genevier, is quite poisonous. It has a dull white pileus and immediately stains a bright yellow, particularly in the base of the stipe. **A. albolutescens** Zeller is also poisonous, at least to some people. It is characterized by an amber yellow stain on the pileus when bruised; the yellow eventually colors the entire cap.

Agaricus meleagris Schaeff.

PILEUS up to 15 cm in diameter; convex, somewhat flattened on top, becoming broadly convex to plane; surface dry; brownish-black on disk, elsewhere covered with small grayish fibrils or squamules. FLESH white, sometimes staining yellow; odor of phenol. GILLS close; grayish-pink, becoming bright pink, and finally dark chocolate-brown. STIPE up to 15 cm long; tapering upward slightly; smooth; white, staining yellow at base; annulus white, superior. SPORES purplish-brown; ellipsoid; smooth; 4-4.5 x 3.5-4 μ. POISONOUS.

A. meleagris is a common species often found in groups or caespitose clumps in western coniferous forests and woodlands. Its grayish fibrillose pileus and strong odor of phenol are good field characters. It has been known to cause rather severe gastrointestinal illness.

Until recently, *A. meleagris* has been mistaken for **A. placomyces**, an edible eastern species. *A. placomyces* is distinguished by the presence of brown droplets on the underside of the unbroken veil.

Agaricus augustus Fr.   Plate 39

PILEUS up to 36 cm in diameter; somewhat cylindrical at first, expanding to convex, and then plane with age; surface dry; yellowish-brown with fine, dark brown scales. FLESH fairly thick; white or yellowish-tinged; lacks odor; taste somewhat almond-like. GILLS close; at first white to pale pink, becoming blackish-brown. STIPE up to 17 cm long; relatively thick; deeply imbedded; yellowish-white; scaly below annulus, smooth above; annulus large, double. SPORES purplish-brown; ellipsoid; smooth; 8-10 x 5-6 μ. EDIBLE and choice.

A. augustus, commonly called the Prince, is a gigantic species and considered a choice edible. It occurs in a variety of habitats from cultivated parkland to coniferous forests, and is usually found in late spring and summer along the Pacific Coast from British Columbia to central California.

A very similar species, **A. subrufescens**, also edible, is distinguished mainly by its smaller spores that measure 6-7.5 x 4-4.5 μ.

FAMILY STROPHARIACEAE

The Strophariaceae contains the genera *Stropharia*, *Naematoloma*, and *Psilocybe*, all of which have purplish-brown or chocolate-brown spores with an apical pore, attached gills, and a filamentous pileus cuticle. They may or may not be annulate, but they never have a volva. Their adnate or adnexed gills differentiate them from the genus *Agaricus* (see above), whose members have free gills.

Stropharia may occur on the ground or on dung, is usually viscid, and has an annulus. *Naematoloma* (listed as *Hypholoma* in older literature) grows on wood and is often caespitose. *Psilocybe* may be dung-inhabiting, but it can also be found on wood chips, in pastures, or in forests; it may or may not have an annulus.

Some Psilocybes are hallucinogenic because they contain certain indoles such as psilocybin and/or psilocin, baeocystin, norbaeocystin, and perhaps other compounds as yet unidentified. Hallucinogenic species often stain blue in some parts, upon bruising or drying. Possession of species containing psilocybin and/or psilocin is illegal.

Most of the Strophariaceae cannot be recommended for eating, because even the few edible western species are not outstanding in flavor, with the exception of *Stropharia rugosoannulata* which is reported to be a choice edible.

Genus *Stropharia*

Stropharia ambigua (Pk.) Zeller

PILEUS up to 10 cm in diameter; ovoid, becoming broadly convex or finally plane; viscid; dull yellow; margin with large, white floccose remnants of veil. FLESH white; thick; lacks a distinctive odor. GILLS adnate; close; grayish to dark purple. STIPE up to 25 cm long; 1-2 cm thick; equal; white;

often with remains of veil, becoming smooth in age. SPORES purplish-brown; ellipsoid; smooth; 11-14 x 6-8 μ. EDIBILITY questionable.

S. ambigua is a tall and handsome mushroom, conspicuous in the fall and winter months along the Pacific Coast. It usually grows gregariously under conifers and alders.

Stropharia aeruginosa (Curt. ex Fr.) Quél.   Plate 40

PILEUS up to 5 cm in diameter; convex to broadly campanulate; smooth; yellow, with a glutinous bluish-green covering at first. FLESH soft; white or somewhat bluish-tinted. GILLS adnate; close; gray, becoming purplish-brown in age. STIPE 5-7 cm long, 4-7 mm thick; viscid; bluish-green; annulus evanescent. SPORES purplish-brown; ovoid; smooth; 7-8 x 4-5 μ. POISONOUS.

S. aeruginosa is found in fall or winter in mixed forests and it is widely distributed. Its green coloration is not common in fungi, but the heavy coat of gluten will often be washed away by rain or disappear upon drying, and the yellow pileus will be exposed.

Stropharia hornemannii (Fr.) Lund. and Nannf.

PILEUS up to 10 cm in diameter; convex to plane, with umbo often present; smooth, but with white floccose patches along margin when young; viscid; reddish-brown to olive-brown. FLESH white to yellowish. GILLS adnate, with decurrent tooth; gray to purplish. STIPE up to 10 cm in length; equal; white, with numerous floccose scales beneath conspicuous, white annulus. SPORES purple-brown; ellipsoid; smooth; 11-13.5 x 5.5-7 μ. EDIBILITY unknown.

S. hornemannii occurs widely over the mountains of the West. It is usually associated with conifers and appears in the fall.

Stropharia semiglobata (Fr.) Quél.

PILEUS up to 4 cm in diameter; hemispheric at first, becoming almost plane; surface viscid or glutinous; bright yellow, soon fading to creamy white. FLESH pale yellow; soft; lacking distinct odor. GILLS adnate; grayish to purple-brown, with edges white. STIPE up to 12 cm long; whitish; viscid; annulus indistinct, soon disappearing. SPORES purplish-brown; ellipsoid; smooth; 15-20 x 8-11 μ. EDIBILITY questionable.

S. semiglobata, a small, dung-inhabiting species, is likely to be found in any stock pasture after the first fall rains.

Stropharia rugosoannulata Farlow ex Murr.

PILEUS 5-15 cm in diameter; hemispheric to convex or plane; smooth or somewhat fibrillose near margin; wine-red or yellowish in age. FLESH thick; firm; white. GILLS adnexed; whitish, becoming grayish-violet to blackish-violet. STIPE 5-8 cm long, 1-1.5 cm thick; whitish; fibrillose; annulus superior; upper surface striate, lower part thick and split into points. SPORES grayish-violet; ovoid to elliptical; smooth; 10-12 x 6-8 μ. EDIBLE and choice.

S. rugosoannulata occurs in damp, cultivated places such as parks, lawns, and gardens, mostly in the fall, or in spring and summer in watered areas in the Pacific Northwest. Because of this habitat, it is believed to have been introduced. This large and striking *Stropharia* is wine-red before the color fades in age. It is said to be a choice edible.

Genus *Naematoloma*

Members of *Naematoloma* have purplish-brown spores, a pileus that may be dry or moist but is never viscid, and lack an annulus. Most, but not all, are densely caespitose and occur on stumps or logs.

Naematoloma capnoides (Fr.) Karst.

PILEUS up to 7 cm in diameter; convex to plane, often with slight umbo; margin inrolled at first; surface smooth, moist; bright orange-yellow to reddish-orange in center, becoming almost white along margin. FLESH white; odor and taste mild. GILLS adnate; white to grayish when young, becoming purplish-brown with age. STIPE up to 10 cm long; moderately thick; whitish above, becoming tan to rusty-brown below. SPORES purplish-brown; ellipsoid; smooth; 6-7.5 x 4-4.5 μ. EDIBLE.

N. capnoides is a common fungus in autumn and winter along the Pacific Coast, often growing in large clusters on stumps and fallen logs. It also occurs in abundance in the Rocky Mountains in fall.

Naematoloma dispersum Karst.

PILEUS up to 4 cm in diameter; convex; smooth; moist; reddish-brown to orange-brown in center, becoming paler toward margin. FLESH whitish to grayish-brown; not brittle; odor and taste not distinctive. GILLS adnate; pallid when young, becoming grayish to purplish-brown, with whitish edges. STIPE up to 10 cm long; slender; equal; tough; marked by lighter and darker zones. SPORES purplish-brown; ellipsoid; smooth; 7-11 x 4-5 μ. EDIBILITY unknown.

N. dispersum is widely distributed during the rainy seasons along the Pacific Coast and in the Rocky Mountains, where it usually grows singly on logs or woody debris in coniferous forests. Its most distinguishing feature is the tall, slender stipe, which is very tough and usually marked by light fibrillose zones along its length.

Naematoloma fasciculare (Fr.) Karst.

PILEUS 2-8 cm in diameter; conic to plane, with margin first inrolled; surface smooth, moist; bright yellow or yellow-green on margin, becoming orange or orange-brown in center. FLESH yellowish, darkening on exposure to air; taste bitter. GILLS adnate; greenish-yellow, becoming olive, finally turning purplish with age. STIPE up to 12 cm long; moderately thick; yellow; fibrillose remains of veil usually present. SPORES purplish-brown; ellipsoid; smooth; 6.5-8 x 3.5-4 μ. POISONOUS.

N. fasciculare is a fungus with which all mushroom hunters soon become familiar. It is poisonous, but its bitter flavor would deter anyone from eating it in any event. The fruiting bodies often occur in large caespitose masses on decaying stumps and logs of conifers and hardwoods.

Genus *Psilocybe*

The Psilocybes are mostly small fungi with a conic or campanulate pileus with attached gills. The stipe is usually thin and long and may or may not have an annulus; spores are purplish-brown or violet-gray. These fungi grow on the ground or on dung, and occasionally on wood.

The genus *Psilocybe* is significant because it contains several hallucinogenic species. They have a bluing reaction in some part upon being bruised or dried. The use of these mushrooms for recreational purposes, particularly in the Pacific Northwest, cannot be ignored. However, possession of those species containing the indoles psilocybin and psilocin is illegal.

KEY TO SPECIES OF *PSILOCYBE*

1a. Growing on dung 2

b. Growing on the ground in pastures, etc.
(rarely on wood) 3

c. Growing on wood in
conifer forests .... *P. pelliculosa*, p. 142

  2a. Known in the West
only from Texas, pileus large,
2-8 cm broad.............. *P. cubensis*, p. 142

  b. Common and widely distributed; pileus
small, 1-2 cm broad ...... *P. coprophila*, p. 142

3a. Caespitose; stipe
10-14 cm long .... *P. strictipes*, p. 144

b. Usually solitary or gregarious;
stipe up to 8 cm long 4

  4a. Ranging from California to British
Columbia; not annulate 5

  b. Known range from Oregon to
British Columbia; superior,
persistent annulus .... *P. stuntzii*, p. 145

5a. Pileus convex, often with wavy margin;
chestnut-brown .... *P. cyanescens*, p. 143

b. Pileus conic to convex, umbonate;
olive-brown.................. *P. baeocystis*, p. 143

c. Pileus acutely conic; dingy-brown to
yellowish when dry.......... *P. semilanceata*, p. 142

Psilocybe pelliculosa (A. H. Smith)
Sing. and A. H. Smith

PILEUS 8-25 mm in diameter; conic to obtusely campanulate; hygrophanous; yellowish-brown to grayish-brown, becoming lighter when dry; stains slightly blue. FLESH thin;
yellowish. GILLS adnate; close; cinnamon-brown to grayish-
brown; edges white or pallid. STIPE 4-8 cm long, 1.5-2.5 mm
thick; equal; pallid above, brownish toward base. SPORES
purplish-brown; ellipsoid; smooth; 8-11 x 4.5-5.5 μ. HAL-
LUCINOGENIC and POISONOUS.

P. pelliculosa is the only forest-inhabiting species in the
genus. A few others may occur on wood chips or stumps in
more grassy, open places such as pastures or lawns. It fruits
in fall and winter in Idaho, the Pacific Coast states, and
British Columbia. Because of the bluing reaction when the
pileus is bruised, it is of course in the suspect group, but *P.
pelliculosa* is only slightly hallucinogenic.

A somewhat similar-appearing mushroom, **P. semilanceata** (Fr. ex Secr.) Kummer, is commonly called Liberty Cap
because of its shape. It is conic with a somewhat knoblike
apex, but it differs by having an inrolled margin and a white
stipe. It is hallucinogenic.

Psilocybe coprophila (Bull. ex Fr.) Kummer

PILEUS 1-2 cm in diameter; convex; dry or viscid; without
pellicle; grayish-brown to dark brown. FLESH thin; brown.
GILLS adnate; far apart; grayish-brown to dark purplish-
brown; edges white. STIPE 2-6 cm long, 1-3 mm thick; equal;
mycelioid base. SPORES purplish-brown; elliptical; smooth;
11-15 x 6.5-9 μ. HALLUCINOGENIC.

P. coprophila is widely distributed wherever dung is
present. This very common little *Psilocybe* is known to
contain small amounts of psilocybin when fresh, although it
does not stain blue. However, bluing has been reported in
the mycelium (Menser, 1977).

Another dung-inhabitant, **P. cubensis** (Earle) Sing., is

native in the Southeastern states, although it has been recorded in the Southwest from Texas. However, it is well known because it is the most common psychotropic mushroom in use in certain circles and can be easily cultivated. As stated before, possession of psilocybin and/or psilocin-containing fungi is illegal. *P. cubensis* is a relatively large species with a conic to convex or umbonate pileus that may be up to 8 cm in diameter. It is whitish or yellowish, becoming pale brown in age, and both the pileus and stipe stain a dark blue; the superior annulus is white when young, but soon becomes colored by the spores.

Psilocybe cyanescens Wakefield   Plate 41

PILEUS 2-5 cm in diameter; convex and often umbonate; margin striate, at times raised and wavy; hygrophanous; chestnut-brown, becoming yellowish; stains blue. FLESH pliable in youth, becoming somewhat brittle. GILLS adnate; rather far apart; reddish-brown with paler edges. STIPE 6-8 cm long, 2.5-6 mm thick; equal or slightly enlarged at base, often with rootlike extensions; whitish, sometimes stains blue. SPORES violet-gray; oblong; smooth; 9-12 x 6.9-9.2 μ. HALLUCINOGENIC.

P. cyanescens ranges from California to British Columbia; it sometimes grows on stumps as well as in pastures or other grassy places. It fruits in the fall and winter.

The chestnut pileus and often wavy margin will distinguish *P. cyanescens* from a somewhat similar species, *P. baeocystis* (see below). Both are known to contain psilocybin and psilocin, as well as other unknown substances, and are considered to be equally potent.

Psilocybe baeocystis Sing. and A. H. Smith

PILEUS 1.5-5 cm in diameter; conic, expanding to convex or plane, umbonate; viscid; olive-brown, becoming tan with brown disc; margin incurved at first, faintly striate, blue or green tints where touched. FLESH pliable; olive-brown or paler. GILLS adnate; close; edges white. STIPE 5-7 cm long, 2-3 mm thick; equal; often sinuate; stuffed with brown pith; yellowish at apex, white below; stains blue. SPORES lavender-gray; oblong; smooth; apical pore; 10-13.3 x 6.3-7 μ. HALLUCINOGENIC.

P. baeocystis (Fig. 18) may be solitary to caespitose on lawns, in gardens, or among wood chips. It occurs in the fall from California to British Columbia.

A somewhat similar species, **P. strictipes** Sing. and A. H. Smith, is much rarer and known only from Oregon northward to British Columbia. It is larger, with a stipe that may be up to 14 cm long, and caespitose. Both of these species stain blue and are believed to be equally high in psilocybin and psilocin content, plus other unidentified substances.

FIG. 18. *Psilocybe baeocystis.*

Psilocybe stuntzii Guzmán and Ott

PILEUS 1-4 cm in diameter; campanulate or convex, becoming plane in age; hygrophanous; dark brown. FLESH fragile, brown. GILLS adnate; far apart; purplish-brown; edges white at first. STIPE 3-8 cm long, 1-5 mm thick; equal; surface powdery; persistent, superior, white annulus. SPORES dark purplish; ellipsoid; smooth; 10.1-12.5 x 6-7.2 μ. HALLUCINOGENIC.

Although just recently described, *P. stuntzii* is said to be the most common member of the genus. It often fruits prolifically in western Washington, occurring also in Oregon and British Columbia. It is also known to contain psilocybin and psilocin, but in lesser amounts than *P. cyanescens*, *P. baeocystis*, and *P. strictipes*.

B. SPORES PINK TO SALMON

The two families of pink- or salmon-spored gilled mushrooms are Rhodophyllaceae and Volvariaceae. There are only a few edibles in either family and several dangerously poisonous species, so this group should be avoided. Identification is often difficult even for experts.

The Rhodophyllaceae are mostly terrestrial (a few Leptonias grow on wood); they have attached gills and angular or furrowed spores, and lack an annulus and volva. The Volvariaceae may occur on soil or wood, or even be parasitic on other fungi. They have free gills and smooth spores, and may or may not be volvate. Species in both families usually have white or grayish gills which eventually become some shade of pink as the spores mature.

FAMILY RHODOPHYLLACEAE

The genera of the Rhodophyllaceae described here are *Clitopilus*, which has longitudinally furrowed spores, and *Entoloma*, *Leptonia*, and *Alboleptonia*, which have angular spores.

Genus *Clitopilus*

Clitopilus prunulus (Scop. ex Fr.) Kumm.

PILEUS 3-10 cm in diameter; convex to depressed; pruinose; margin inrolled, later irregularly elevated or wavy; pale gray. FLESH white; soft; odor and taste farinaceous. GILLS decurrent; at first white, becoming pink. STIPE 3-6 cm long, 7-20 mm thick; tapering downward; whitish; base with white tomentum. SPORES pale salmon-pink; fusiform; ridged; 11-14 x 4-6 μ. EDIBLE.

C. prunulus is a common species which occurs in both deciduous and coniferous forests. Its mealy taste and odor are reported to disappear when cooked.

Genus *Entoloma*

Entoloma contains mostly large, fleshy species; many are highly poisonous. They usually range from white to shades of gray or brown; and the attached gills are at first white, grayish, or yellowish.

Entoloma lividum Fr.   Plate 42

PILEUS 7-15 cm in diameter; convex to campanulate, in age nearly plane, with a slight umbo; margin at first inrolled and striate; viscid in wet weather; grayish-yellow to pale grayish-brown. FLESH firm; thick; white. GILLS adnate-sinuate; broad; whitish to pale gray, becoming salmon-pink. STIPE 4-12 cm long, 1-3 cm thick; equal; hollow; whitish or grayish. SPORES salmon-pink; angular; apiculate; with one large oil drop; 7-10 x 7-9 μ. POISONOUS.

E. lividum is a widely distributed and extremely poisonous species which has caused deaths. It occurs under both conifers and hardwoods.

E. rhodopolium (Fr.) Kumm. is also poisonous and is found in mixed woods. It is smaller, 4-8 cm broad, a pale livid gray, shiny when dry, with whitish eroded gills and a pure white stipe. The spores are subglobose and angular, 6-9 μ. E.

clypeatum (L. ex Fr.) Kumm. is a moderately large mushroom, up to 10 cm broad, which occurs infrequently in
California under conifers or hardwoods, and is also poisonous. It has a hygrophanous brownish or brownish-gray fibrillose pileus and whitish serrate gills; its spores are subglobose
and angular, 7-9.5 x 6-7.5 μ. **E. madidum** (Fr.) Gill., a fairly
large species with a blue-gray pileus, is common in fall and
winter along the Pacific Coast; its edibility is questionable.

Genus *Leptonia*

The West has an abundance of Leptonias in the fall and
winter mushroom season, particularly along the coast. They
are relatively small fungi, usually in shades of brown or
bluish-black, with a pileus averaging 1-5 cm in diameter.
Since many are hygrophanous, there may be a noticeable
change in color as the carpophores age. These delicate and
very attractive little mushrooms are almost always found on
the forest floor, although a few species are lignicolous.

Some of the darker species, such as **L. corvina** (Kühn.)
P. D. Orton and **L. decolorans** (Horak) Largent, are especially striking, with a steel-blue pileus and stipe. Edibility is
unknown. The Pacific Coast species of *Leptonia* have been
monographed recently (Largent, 1977) in an extremely
thorough work.

Genus *Alboleptonia*

Alboleptonia is very similar to the genus *Leptonia* (see
above), but the pileus in its members is white to pale ashy with
some microscopic differences in the surface tissue.

A. sericella Largent and Benedict is pure white in all parts,
with a shining, silky, or minutely squamulose pileus. Its
spores are elongate, with a prominent apiculus, and measure
9-13 x 6-8 μ. In spite of its small stature, 2-5 cm tall, it is
conspicuous; we have found it growing gregariously in Coast
Redwood forests in northern California. Its edibility is
unknown.

FAMILY VOLVARIACEAE

The family Volvariaceae differs from the one other pink-spored family, the Rhodophyllaceae (see above), by having nonangular, elliptical, smooth spores and free gills. The genus *Volvariella* always has a volva but no annulus; the genus *Pluteus* lacks both volva and annulus; and a rare genus, **Chamaeota**, has an annulus but no volva. All may occur on soil or wood and, in a few unusual cases, they are parasitic on other fungi.

Genus *Volvariella*

Volvariella speciosa var. **speciosa** (Fr.) Sing.

PILEUS 5-10 cm in diameter; ovate, becoming expanded to flat, sometimes with slight umbo; viscid; smooth; margin not striate; white or slightly grayish-tinged. FLESH soft; thin; white. GILLS free; broad; crowded; white. STIPE 10-20 cm long, 1-2 cm thick; equal; white; volva large, splitting at top, white. SPORES pink; ellipsoid; smooth; 14-20 x 9-12 μ. EDIBILITY questionable; see below.

V. speciosa var. *speciosa* is found on rich soil or in leafmold in woodlands, and we have seen prolific fruitings in cultivated fields in northern California where the ground was literally covered with thousands of these mushrooms.

There have been conflicting reports about the edibility of *V. speciosa*, but var. *speciosa* is commonly eaten by many Californians and considered to be very good. European literature usually lists it as poisonous, however, and even responsible for some deaths. Perhaps this problem is due to **V. speciosa** var. **gloiocephala** (D.C. ex Fr.) Sing., which differs by having a much darker brownish-gray pileus with a striate margin and smaller spores, 11-13 x 6-7.5 μ. We have not seen var. *gloiocephala* but it has been recorded as occurring in Berkeley, California, in February.

Volvariella surrecta (Knapp) Sing.
Volvaria loveiana (Berk.) Gill.

PILEUS 2-5 cm in diameter; ovate at first, becoming campanulate to convex; fibrillose; dry; yellowish-white. FLESH thick; firm; white. GILLS free; crowded; whitish. STIPE 1-4 cm long, 5-7 mm thick; fibrillose; yellowish-white; volva large; usually lobed; concolorous with stipe. SPORES pink; ovoid; 5.4-7.6 x 3.4-4.9 M. EDIBILITY questionable; see below.

V. surrecta is a rare fungus that is parasitic on other gilled mushrooms, usually *Clitocybe nebularis*, causing the pileus of its host to become distorted. Though it is said to be edible, its small size and rarity would seem to make it of little importance as food.

Genus *Pluteus*

Pluteus is the only genus of the family Volvariaceae that lacks both annulus and volva. It has a central fleshy or slightly cartilaginous stipe and pink globose or subglobose spores. Most species are of soft consistency and tend to decay in a short time.

Pluteus cervinus (Fr.) Quél.

PILEUS 5-14 cm in diameter; convex to campanulate or plane and umbonate; smooth or slightly fibrillose; moist; dark brown, becoming pale tan; often darker on disc. FLESH soft; thick; white; slight odor and taste of radish. GILLS free; broad; crowded; whitish; large pleurocystidia, with hornlike projections. STIPE 5-12 cm long, 6-15 mm thick; base enlarged and slightly myceloid; pallid to tan. SPORES pale pink; ellipsoid; smooth; 5-5.7 x 4-6 μ. EDIBLE, but poor; see below.

P. cervinus is the most common member of the genus and is cosmopolitan in the temperate zone. It is often found in abundance on both hardwoods and conifers, particularly on decaying logs or sawdust piles. It is commonly known as the Deer Mushroom because of its inconspicuous light and dark coloration, similar to the spots which camouflage a fawn. The pleurocystidia (found on the face of the gills) are large, with hornlike projections. It is edible, and the radish flavor is said to disappear in cooking; but the flavor is unremarkable, according to those who have eaten it.

P. umbrosus (Pers. ex Fr.) Kumm., which occurs on decayed conifer wood along the Pacific Coast, is similar to *P. cervinus* but has a blackish-brown pileus and gills with the edges minutely fringed and smoky-brown. This coloration is caused by dark cells in the cheilocystidia (cystidia on the gill edges). **P. magnus** McClatchie has a blackish, wrinkled pileus and a shorter, more robust stipe, and has been recorded in California. Both of these species are said to be edible.

C. SPORES OCHRE, CLAY-COLOR, OR BROWN

There are three families in this yellow- to brown-spored group. Very few species are edible, and some are extremely poisonous. Most members of the family Bolbitiaceae are small and grow in grassy areas or on dung. The Paxillaceae is a very small family containing fairly large, fleshy species with decurrent gills. The Cortinariaceae is a large family, most of whose members have a cortina or weblike veil; it is estimated that there may be as many as 600 species in the genus *Cortinarius* alone. Few are edible and some, such as *Galerina autumnalis*, are deadly.

FAMILY BOLBITIACEAE

The Bolbitiaceae contains three genera with yellow-brown, rusty-brown, or dark brown spores that may be smooth or ornamented, but always have a germ pore at the apex. They are related to the Coprinaceae; and members of the genera *Conocybe* and *Bolbitius* often resemble this dark-spored family in their fragile, plicate pileus; free or almost free gills; and long, thin stipe. The third genus, *Agrocybe*, is more robust and fleshy, with adnate gills.

One extremely poisonous species, **Conocybe filaris** Fr., occurs on lawns in California in the fall and winter. It has a brown campanulate pileus, 5-15 mm in diameter, and a very slender, fragile stipe that may be 2-4 cm long, with a conspicuous annulus. The spores are rusty-brown, elliptical, smooth with a germ pore at the apex, 7.5-13.2 x 3.5-6.5 μ. It contains amanitin, the same toxin that is present in the deadly *Amanita phalloides*.

Genus *Bolbitius*

Members of *Bolbitius* are rusty-spored fungi with a viscid, striate pileus and gills that tend to deliquesce somewhat, as in species of *Coprinus*.

Bolbitius vitellinus Fr.

PILEUS up to 5 cm in diameter; conic to convex, becoming plane with age; surface smooth, viscid; margin striate; egg-yellow to orange-brown. FLESH yellowish; soft; lacks distinct odor. GILLS narrowly attached to stipe; pallid to yellowish; somewhat deliquescing with age. STIPE up to 10 cm long; slender; hollow; pale yellow with white, floccose covering. SPORES rusty-brown; ellipsoid, with one end truncate; smooth; 12-15 x 7-8.5 μ. EDIBLE, but poor.

B. vitellinus is a common species in meadows and pastures, where it grows on dung, or where there is thick grass. It appears early after the start of the fall rains as well as in spring in some areas. The habitat and the bright yellow, viscid pileus make it conspicuous. It is not poisonous, but its fragile consistency makes it worthless for eating.

Genus *Agrocybe*

Species of *Agrocybe* are usually found on the ground in grassy woodlands. The carpophores are much fleshier than those of *Conocybe* and *Bolbitius*, and they may or may not have an annulus. The spores are dark brown with a germ pore.

Agrocybe praecox Fr.   Plate 43

PILEUS 2-6 cm in diameter; convex with incurved margin at first, becoming nearly plane; smooth; dry; pallid to yellowish or light tan. FLESH moderately thick, white. GILLS adnate to adnexed; whitish to pale brown in age. STIPE 3-8 cm long, 4-8 mm thick; equal or slightly bulbous; white; annulus persistent, white. SPORES dark brown; ellipsoid; smooth, with a germ pore; 8-13 x 5.5-7 μ. EDIBLE.

A. praecox is a very common springtime mushroom which occurs singly or scattered in woodlands or sometimes on lawns in the western states. It is edible, but should be carefully distinguished from some Hebelomas (see below) which are poisonous.

FAMILY PAXILLACEAE

Members of Paxillaceae are fleshy fungi that occur on soil, forest debris, or rotton wood. They may or may not be stipitate; and the stipe, if present, may be central or eccentric. The gills are decurrent and easily separable from the pileus. Spores range from chocolate-brown to clay-color; are ovoid, ellipsoid, or oblong; smooth; and lack an apical pore.

The genus *Paxillus* was formerly regarded as the single member of the family, but now many authorities have transferred *Phylloporus* to the Paxillaceae.

Pl. 1 *Chlorociboria
aeruginascens* (p. 24)

Pl. 2 *Sarcoscypha coccinea* (p. 24)

Pl. 3 *Helvella
elastica* (p. 30)

Pl. 4 *Gyromitra
californica* (p. 32)

Pl. 6 *Gyromitra infula* (p. 33)

Pl. 5 *Gyromitra esculenta* (p. 34)

Pl. 7 *Morchella angusticeps* (p. 37)

Pl. 8 *Caloscypha fulgens* (p. 38)

Pl. 9 *Trichoglossum hirsutum* (p. 41)

Pl. 10 *Hypomyces lactifluorum* (p. 44)

Pl. 11 *Pseudohydnum gelatinosum* (p. 48)

Pl. 12 *Phlogiotis helvelloides* (p. 49)

Pl. 13 *Auricularia auricula* (p. 50)

Pl. 14 *Dentinum repandum* (p. 53)

Pl. 15 *Hydnellum suaveolens* (p. 55)

Pl. 16 *Hydnellum peckii* (p. 56)

Pl. 17 *Hericium erinaceus* (p. 56)

Pl. 18 *Clavariadelphus*
borealis (p. 62)

Pl. 19 *Ramaria formosa* (p. 67)

Pl. 20 *Ramaria stricta* (p. 68)

Pl. 21 *Sparassis radicata* (p. 69)

Pl. 22 *Gomphus clavatus* (p. 71)

Pl. 23 *Gomphus floccosus* (p. 71)

Pl. 24 *Cantharellus cibarius* (p. 72)

Pl. 25 *Ganoderma oregonense* (p. 77)

Pl. 26 *Lenzites betulina* (p. 81)

Pl. 27 *Leccinum scabrum* (p. 87)

Pl. 28 *Suillus cavipes* (p. 91)

Pl. 29 *Suillus grevillei* (p. 92)   Pl. 30 *Suillus pungens* (p. 94)

Pl. 31 *Suillus subolivaceus* (p. 95)

Pl. 32 *Suillus
tomentosus* (p. 96)

Pl. 33 *Boletus pulcherrimus* (p. 103)

Pl. 34 *Boletus regius* (p. 105)

Pl. 35 *Boletus zelleri* (p. 107)

Pl. 36 *Calvatia
sculpta* (p. 113)

Pl. 37 *Phallus impudicus* (p. 122)

Pl. 38 *Agaricus subrutilescens* (p. 132)

Pl. 39 *Agaricus augustus* (p. 135)

Pl. 40 *Stropharia aeruginosa* (p. 137)

Pl. 41 *Psilocybe cyanescens* (p. 143)

Pl. 42 *Entoloma lividum* (p. 146)

Pl. 43 *Agrocybe praecox* (p. 152)

Pl. 44 *Paxillus atrotomentosus* (p. 154)

Pl. 45 *Heboloma
crustuliniforme* (p. 158)

Pl. 46 *Inocybe geophylla*
var. *lilacina* (p. 160)

Pl. 47 *Galerina autumnalis* (p. 162)

Pl. 48 *Cortinarius
subfoetidus* (p. 166)

Pl. 49 *Cortinarius
californicus* (p. 167)

Pl. 50 *Cortinarius phoeniceus* var. *occidentalis* (p. 167)

Pl. 51 *Cortinarius sanguineus* (p. 167)

Pl. 52 *Panaeolina foenisecii* (p. 170)

Pl. 54 *Coprinus micaceus* (p. 176)

Pl. 53 *Panaeolus campanulatus* (p. 173)

Pl. 55 *Coprinus atramentarius* (p. 177)

Pl. 57 *Chroogomphus rutilus* (p. 182)

Pl. 56 *Gomphidius subroseus* (p. 180)

Pl. 58 *Russula densifolia* (p. 186)

Pl. 59 *Russula sanguinea* (p. 187)

Pl. 60 *Russula
xerampelina*
(p. 188)

Pl. 61 *Lactarius rufus* (p. 191)

Pl. 62 *Lactarius aurantiacus* (p. 192)

Pl. 63 *Lactarius uvidus* (p. 194)

Pl. 64 *Lactarius deliciosus* (p. 195)

Pl. 65 *Amanita phalloides* (p. 198)   Pl. 66 *Amanita rubescens* (p. 201)

Pl. 67 *Amanita pantherina* (p. 203)

Pl. 68 *Amanita muscaria* (p. 203)

Pl. 69 *Amanita calyptroderma* (p. 205)

Pl. 70 *Hygrophorus
coccineus* (p. 211)

Pl. 71 *Hygrophorus conicus* (p. 212)

Pl. 72 *Hygrophorus
speciosus* (p. 213)

Pl. 73 *Hygrophorus
camarophyllus* (p. 214)

Pl. 74 *Hygrophorus bakerensis* (p. 215)

Pl. 75 *Hygrophorus*
olivaceoalbus (p. 215)

Pl. 76 *Chlorophyllum*
molybdites (p. 219)

Pl. 77 *Lepiota naucina* (p. 221)

Pl. 78 *Lepiota clypeolaria* (p. 223)

Pl. 79 *Lepiota flammeatincta* (p. 224)

Pl. 80 *Lepiota rubrotincta* (p. 224)

Pl. 81 *Asterophora lycoperdoides* (p. 227)

Pl. 82 *Cystoderma fallax* (p. 228)

Pl. 83 *Mycena lilacifolia* (p. 236)

Pl. 84 *Mycena purpureofusca* (p. 238)

Pl. 85 *Clitocybe odora* (p. 243)

Pl. 86 *Laccaria laccata* (p. 247)   Pl. 87 *Lyophyllum montanum* (p. 248)

Pl. 88 *Collybia familia* (p. 250)

. 90 *Pleurotus porrigens* (p. 253)   Pl. 89 *Collybia umbonata* (p. 250)

. 92 *Lepista nuda* (p. 259)   Pl. 91 *Flammulina velutipes* (p. 257)

Pl. 93 *Tricholoma pardinum* (p. 262)

Pl. 94 *Tricholoma terreum* (p. 263)

Pl. 95 *Tricholoma vaccinum* (p. 263)

Pl. 96 *Tricholoma virgatum* (p. 264)

Genus *Phylloporus*

Phylloporus is a perplexing genus which was formerly classified in the Boletaceae (see Chapter 3 above) because it had similarly shaped spores and gills which stain blue, as do the tubes of many boletes. It appears to form a link between the boletes and the agarics.

Phylloporus rhodoxanthus (Schw.) Bres.

PILEUS 2-9 cm in diameter; convex to expanded and depressed; dry; minutely tomentose; dull reddish-brown, or ochraceous to olivaceous. FLESH firm, thick, pale yellow. GILLS decurrent; thick; narrow; sometimes anastomosing or appearing slightly poroid; bright yellow, faintly bruising blue or dingy-brown. STIPE 2-6 cm long, 5-20 mm thick; tapering downward; buff or reddish-yellow with small reddish-brown scales. SPORES yellowish-brown; elliptical to almost fusiform; smooth; 11-15 x 4-5.6 μ. EDIBLE and choice.

P. rhodoxanthus occurs under both conifers and hardwoods in the fall in the Northwest. It will probably have to be separated into several varieties, because of different color forms.

Genus *Paxillus*

Though *Paxillus* is a very small genus, it is widely distributed and its species easily recognizable. *P. involutus* is to be noted because it can be poisonous under certain circumstances.

Paxillus involutus (Batsch ex Fr.) Fr.

PILEUS 4-12 cm in diameter; convex to depressed in age; sometimes subviscid; slightly tomentose; yellowish-brown, reddish-brown or olivaceous-brown; margin striate, persistently inrolled until aged. FLESH thick; pale yellowish; stains brown. GILLS decurrent, sometimes anastomosing on the stipe; close; yellow at first, staining dull brown when bruised. STIPE 4-6 cm long, 1-2.5 cm thick; central or eccentric; smooth; solid; somewhat enlarged at base; dirty yellowish-brown or concolorous with pileus. SPORES reddish-ochraceous; elliptical; smooth; 7-9 x 4-6 μ. POISONOUS.

P. involutus derives its name from the conspicuously inrolled (involute) margin of the pileus. This rather dingy and unattractive mushroom is always associated with conifers, and sometimes fruits abundantly in fall and winter in the West. Although listed as edible in most older books, it is now known to be toxic to some people who eat it repeatedly, because there can be a cumulative effect.

Paxillus atrotomentosus (Batsch ex Fr.) Fr.

Plate 44

PILEUS 5-12 cm in diameter; convex to plane or depressed; tomentose; dry; margin inrolled at first; tan to rusty-brown or dark brown. FLESH thick; spongy or firm; white. GILLS decurrent; narrow; close; forked and often anastomosing; yellowish to yellow-brown. STIPE 3-12 cm long, 1-3 cm thick; straight or curved; usually eccentric; brown and densely hairy. SPORES yellowish; oval; smooth; 5-6 x 3-4 μ. EDIBLE, but poor.

P. atrotomentosus occurs on rotting conifer wood in fall and winter and is common and widely distributed. It is reported to be edible but of poor consistency.

P. panuoides Fr. is smaller, with a pileus from 3-8 cm in diameter, and lacks a stipe. It is also lighter, with pale yellow or yellow-orange gills, and usually occurs on conifer wood but is occasionally found on hardwoods. Its edibility is unknown.

FAMILY CORTINARIACEAE

The Cortinariaceae is a very large family, which contains some very beautiful mushrooms and also some extremely poisonous ones. Certain species of *Galerina* contain the same toxins as some of the deadly Amanitas.

The genus *Cortinarius* has the distinction of possessing the greatest number of species of gilled mushrooms. Because of the presence of a *cortina*, a cobweb-like veil that at first covers the young gills, the genus is usually easily recognized, but identification of most of the species is difficult even for expert mycologists.

Members of the Cortinariaceae are stipitate (with the sole exception of *Crepidotus*, which is sessile), the gills are attached, there may or may not be an annulus, and a volva is lacking except in the genus *Rozites*, which has some inconspicuous remains of a second veil at the base. The spores are various shades of brown, rust, clay-color, or ochre, and they may be smooth or ornamented.

KEY TO GENERA OF CORTINARIACEAE

1a. Pileus dry and somewhat wrinkled;
   membranous annulus; remains of second
   veil at base . . . . *Rozites*, p. 156
  b. Pileus dry, moist, or viscid;
   cobweb-like veil . . . . *Cortinarius*, p. 163
  c. Not as above 2
     2a. Typically lignicolous; spores
        bright rusty-brown or
        orange-brown . . . . *Gymnopilus*, p. 156
       b. Lignicolous or terrestrial;
        spores not as above 3
3a. Spores dark brown or
   grayish-brown . . . . *Pholiota*, p. 157
  b. Spores brownish-yellow to clay-color;
   pileus viscid . . . . *Hebeloma*, p. 158

c. Spores brown or reddish-brown **4**
   4a. Pileus usually moist; on wood
       or among moss .... *Galerina*, p. 161
    b. Pileus usually dry and fibrillose;
       terrestrial .... *Inocybe*, p. 159

Genus *Rozites*

Members of *Rozites* can be differentiated from other Cortinariaceae by the membranous double veil. There is only one common species in North America.

Rozites caperata (Pers. ex Fr.) Karst.

PILEUS 5-10 cm in diameter; ovoid, becoming broadly convex; dry; surface irregularly wrinkled; pale orange, with a white pruinose covering in youth; margin paler, inrolled at first. FLESH firm; white. GILLS adnate to adnexed; close; pallid, becoming rusty-brown. STIPE 5-10 cm long, 1-2 cm thick; equal or enlarged at base; pallid; annulus membranous, superior, white. SPORES rusty-brown; elliptical; ornamented; 12-14 x 7-9 μ. EDIBLE and choice.

R. caperata is called the Gypsy Mushroom.

Genus *Gymnopilus*

The lignicolous habitat and bright rusty-brown or orange-brown spores make *Gymnopilus* an easily recognized genus. Some species are extremely bitter and none is known to be desirable for food, even if harmless.

G. spectabilis (Fr.) Sing. is a very large, yellow to orange fungus that is probably familiar to most collectors in conifer and hardwood forests. It has been suspected of being hallucinogenic, but its toxic properties are as yet undetermined. It is hard to see how anyone could possibly eat it, however, because it is extremely bitter.

Genus *Pholiota*

Most Pholiotas occur on wood and are rather large and conspicuous. They have attached gills; an annulus; and yellow-brown to rusty-brown, smooth spores. They are usually scaly and often occur in very large caespitose clusters.

One species that is not lignicolous is **P. terrestris** Overholts, which is dry and has numerous reddish-brown scales on the pileus and stipe. Though edible, it is of poor quality.

There are some good edible species of *Pholiota*, and they can often be collected in quantity. However, **P. aurivella** (Fr.) Kumm., a rather robust and striking fungus, with purplish-red scales on an orange background, is suspected of containing gastrointestinal irritants. *P. squarrosa* (see below) is known to cause digestive disturbance.

Pholiota mutabilis (Fr.) Quél.

PILEUS 1.5-6 cm in diameter; bluntly conic to campanulate or plane, often with obtuse umbo; smooth or minutely fibrillose; viscid; hygrophanous; dark reddish-brown to clay-color; fading on disc or in area around disc; margin striate. FLESH thin; watery; pallid. GILLS adnate to decurrent; crowded; whitish, becoming yellowish or dull cinnamon. STIPE 4-10 cm long, 2-12 mm thick; tapering downward; becoming hollow; pallid or clay-color above the annulus, dark brown and densely scaly below. SPORES deep rusty-brown; subovoid; smooth; 5.5-7.5 x 3.7-4.5 μ. EDIBLE.

P. mutabilis is called the Changing Pholiota because of its hygrophanous character. It is found frequently in densely caespitose clusters on decaying wood in the Pacific Northwest and the Rocky Mountains. Although edible and said to be rather good, it should be identified with certainty, because it bears great superficial resemblance to a deadly poisonous mushroom, *Galerina autumnalis* (see below).

Pholiota squarrosa (Fr.) Kumm.

PILEUS 3-12 cm in diameter; convex, becoming campanulate; dry; coarse yellowish-brown scales on ochraceous background. FLESH thick; pale yellow; taste mild or rancid in age. GILLS adnate; crowded; pale yellow or greenish, becoming rusty-brown. STIPE 4-10 cm long, 4-12 mm thick; equal or tapering downward; solid; covered with yellowish-brown scales up to superior annulus. SPORES brown; elliptical; smooth; with apical pore; 5-8 x 3.5-4.5 μ. EDIBILITY questionable; see below.

P. squarrosa occurs in densely caespitose clumps on wood. There are conflicting reports about its edibility; it has been reported to cause gastrointestinal illness in some people, and is also said to have a rancid flavor.

A somewhat similar species, **P. squarrosoides** (Pk.) Sacc., found in similar habitats, has rusty-brown, pointed scales that sometimes cover the entire pileus. It is considered a choice edible.

Genus *Hebeloma*

Hebeloma is a difficult and poorly studied genus but, since it contains some very poisonous species and no known edibles, the mushroom hunter should be aware of its characteristics. Species of *Hebeloma* grow on the ground, and many have a strong odor. The pileus is convex and always viscid, and the stipe is dry. The gills are adnexed to emarginate, with long cylindric or club-shaped cheilocystidia. There is no annulus, but a few species have a fibrillose veil. The spores are dark yellow-brown to clay-color, and smooth or minutely warted.

Hebeloma crustuliniforme (Bull. ex St. Amans) Quél.
Plate 45

PILEUS 3-7 cm in diameter; convex, expanding to broadly convex or plane with broad umbo; viscid; smooth; cream-color with brownish umbo; margin inrolled, becoming raised. FLESH thick; white; odor of radish. GILLS adnexed; close; white, becoming dull brown; edges crenulate. STIPE 4-8 cm long, 6-10 mm thick; solid; equal or bulbous at base; white; abundant white mycelium. SPORES yellow-brown to dull brown; broadly elliptical to almond-shaped; minutely punctate; 9-11.5 x 5-5.7 μ. POISONOUS.

H. crustuliniforme is a very common species on the Pacific Coast and in the northern Rocky Mountains. It fruits in the fall and winter in coniferous or mixed forests.

A larger, related species, **H. sinapizans** (Paulet ex Fr.) Gill., has a dull vinaceous-brown pileus and a distinctly scaly stipe. **H. mesophaeum** (Pers. ex Fr.) Quél. is smaller, with a brown pileus not over 4 cm in diameter, and it has a fibrillose veil. Both have the same radish odor as *H. crustuliniforme*. All three species are poisonous, but their toxins have not yet been determined.

Genus *Inocybe*

Members of *Inocybe* are another extremely suspect group, because most that have been studied so far have been found to contain muscarine. Since it seems highly probable that all species are poisonous, they should be avoided.

Inocybe has been estimated to have about 400 species in North America. They are small fungi, which usually grow on the ground in woodlands. The pileus is conic to convex or campanulate, usually dry, fibrillose, and sometimes radially split. The gills are adnate to adnexed, and there is no annulus or volva, but sometimes a fibrillose veil is present. Spores are some shade of brown and may be smooth or tuberculate.

Nearly all have a characteristic odor, usually unpleasant; but one of the rare exceptions is **I. pyriodora** (Pers. ex Fr.) Kumm., which smells like ripe pears. The flesh of some species will stain red when bruised.

Inocybe geophylla var. geophylla (Sow. ex Fr.) Kumm.

PILEUS 1.5-3 cm in diameter; conic to campanulate, umbonate; fibrillose; dry; white. FLESH thick on disc; white; odor unpleasant. GILLS adnate to adnexed; close; whitish, becoming dull brown. STIPE 2-6 cm long, 5-8 mm thick; equal; white; sometimes with remains of fibrillose veil. SPORES dull brown; elliptical; smooth; 7-10 x 4.5-6 μ. POISONOUS.

I. geophylla var. *geophylla* is very common throughout North America. A delicately lilac-tinted relative is **I. geophylla** var. **lilacina** (Pk.) Boudier (Plate 46). Both occur under conifers or hardwoods, sometimes intermixed in the locality; both are poisonous.

Inocybe sororia Kauf.

PILEUS 3-6 cm in diameter; conic, becoming campanulate with an acutely pointed umbo; with long radial cracks through the cuticle; yellowish to dingy-olivaceous in age.

FIG. 19. *Inocybe sororia.*

FLESH thin; pallid to olivaceous; odor of green corn. GILLS adnexed; crowded; whitish to yellowish, becoming brown. STIPE 5-10 cm long, 2-8 mm thick; equal or slightly bulbous; fibrillose; sometimes twisted; concolorous with pileus, or whitish on lower part. SPORES dull brown; elliptical; smooth; 10-13 x 5.5-7 μ. POISONOUS.

I. sororia (Fig. 19) has a strong odor of green corn which makes it easily recognizable. It is abundant on the Pacific Coast. The muscarine content in this species is high.

Inocybe godeyi Gill.

PILEUS 3-5 cm in diameter; conic to campanulate, with an obtuse umbo; silky-fibrillose; becoming cracked in age; yellow to irregularly reddish or reddish-brown. FLESH firm; white; reddish under cuticle; stains red where cut; odor spermatic. GILLS adnexed; crowded; whitish, becoming reddish or rusty-brown spotted; edges white, floccose. STIPE 2.5-7 cm long; 6-10 mm thick; equal with abruptly bulbous base; fibrillose; concolorous with pileus; sometimes white at apex. SPORES olivaceous brown; almond-shaped; smooth; 10-15 x 6-7.5 μ. POISONOUS.

I. godeyi occurs in California under conifers. It has a low muscarine content, but it could be toxic if much was consumed. In any event, its unpleasant odor would probably cause it to be shunned for the table.

Genus *Galerina*

Galerina is an extremely important genus for the mushroom hunter to master, because some species contain the same toxins and are as deadly as some of the most poisonous Amanitas.

Species of *Galerina* are typically small, hygrophanous mushrooms with attached gills and a slender, brittle stipe. They usually range from pallid to tawny or deep russet hues. The spores are cinnamon, clay-color, or very dark brown, roughened or finely warted. They often occur on wood, and when they grow on the ground they have a particular affinity for mossy habitats.

Galerina autumnalis (Pk.) A.H. Smith and Sing.

Plate 47

PILEUS 2.5-5 cm in diameter; convex, sometimes with low umbo; smooth; viscid; yellowish-brown when wet, pale buff when dry; margin slightly striate. FLESH thin; watery-brown; odor farinaceous, or sometimes slightly like cucumber. GILLS adnate; close; pale brown, becoming rusty-brown. STIPE 1.5-6 cm long, 3-7 mm thick; hollow; equal; fibrillose; pale brown, darker at base; superior white ring left by hairy veil. SPORES rusty-brown; elliptical; wrinkled; 8.5-10 x 5-6.5 μ. POISONOUS.

G. autumnalis can be just as fatally poisonous as *Amanita phalloides* (see below) and other toxic Amanitas. As its name indicates, it usually fruits in the fall, only rarely in the spring. It is common and widely distributed in moist forests; it may be solitary, scattered, or in densely caespitose clumps on decayed conifer or hardwood logs. it even occurs on the tundra in northern Alaska. There is a possibility of its being mistaken for the edible *Pholiota mutabilis* (see above), but microscopic examination of the spores would show smooth spores in *Pholiota*.

Two other deadly poisonous species of *Galerina* have a strong resemblance to *G. autumnalis*. **G. marginata** (Batsch. ex Secr.) Kühn. occurs also on decaying wood. It has a smooth, hygrophanous, cinnamon-brown pileus that fades to yellowish-orange or yellowish-buff when dry. **G. venenata** A.H. Smith is particularly dangerous, because it occurs in grass, sometimes in lawns, where children have been seriously poisoned by it. It is known to occur only in the Pacific Northwest, and fortunately is rare. The pileus is smaller than that of *G. autumnalis*, 1-3.5 cm in diameter, cinnamon-brown and moist. The gills are yellowish-brown to dull cinnamon. There is a superior ring left by a hairy veil, and white mycelium at the base of the stipe. Spores are rusty-brown, elliptical, roughened, 8-11 x 6-6.5 μ.

Genus *Cortinarius*

Cortinarius (Fig. 20) is the genus containing the largest number of gilled fungi, estimated possibly to number up to 600 species; it is therefore not surprising that such a vast group has not been thoroughly studied in North America. Although the genus can be readily recognized by the *cortina*, a cobweb-like veil that at first covers the gills, most of the species are quite difficult to identify. They are mostly medium-sized to large fleshy mushrooms with a convex pileus, attached gills, a stipe that does not readily separate from

FIG. 20. *Cortinarius* sp.

the pileus, and rusty-brown roughened spores. The cortina may soon disappear or, if substantial enough, remain on the stipe as a hairy annulus or adhere to the margin of the pileus. There may also sometimes be an outer veil that will eventually leave ringlike zones around the lower stipe. *Cortinarius* can be distinguished from a similar genus, *Hebeloma* (see above) by the absence of *cheilocystidia* (sterile cells) on the gill edges.

Species of *Cortinarius* usually grow on the ground, particularly in mossy habitats. Only a very few occur on rotting wood. There are some that cause gastrointestinal disturbance, and none should be eaten. However, many of them have beautiful coloration and should be appreciated for that alone.

KEY TO SOME SPECIES OF *CORTINARIUS*

A. Gills shades of purple, violet, or lilac
  1a. Pileus, gills, and stipe
    dark violet .... *C. violaceus*, p. 165
   b. Not as above 2
    2a. Pileus pale lilac; gills
     dull violet; odor
     unpleasant .... *C. traganus*, p. 165
     b. Pileus, gills, and stipe bright
     lilac; odor sweet and
     nauseous .... *C. subfoetidus*, p. 166
B. Gills red or orange-red
  1a. Pileus and stipe concolorous 2
   b. Not as above 3
    2a. Pileus and stipe
     blood-red .... *C. sanguineus*, p. 167
     b. Pileus and stipe cinnabar-
     red .... *C. cinnabarinus*, p. 166
  3a. Pileus rusty-red; stipe
    dull-orange .... *C. californicus*, p. 167
   b. Not as above 4

Cortinarius violaceus (Fr.) S. F. Gray

PILEUS 5-12 cm in diameter; convex, sometimes with slight umbo; dry; erect tufts of hairs over disc; hairs more appressed toward margin; dark violet, with a metallic sheen in age. FLESH thick; firm; dark violet to grayish-violet in age. GILLS adnate; rather far apart; dark violet. STIPE 7-12 cm long, 1-2.5 cm thick above, usually clavate below; fibrillose; dark violet. SPORES rusty-brown; ellipsoid; roughened; 13-17 x 8-10 μ. POISONOUS.

C. violaceus is deep violet in all parts. It fruits in the fall in conifer forests along the Pacific Coast and in the northern Rocky Mountains.

C. purpurascens Fr. is another purple species, but its color is much lighter and its flesh and gills change to deep purple when bruised. It can cause gastrointestinal trouble when eaten raw.

Cortinarius traganus (Weinm. ex Fr.) Fr.

PILEUS 5-13 cm in diameter; convex, becoming almost plane or with broad umbo; dry; fibrillose; pale lilac, soon fading to tan; margin at first inrolled, later often splitting. FLESH thick; firm; yellowish to yellowish-brown; odor pungent, taste bitter. GILLS adnate; rather far apart; dull violet, becoming brown; edges paler and crenulate. STIPE 5-9 cm long, 1.5-4 cm thick; stuffed; bulbous; lilac, sometimes pallid below with whitish zones; cortina pale violet. SPORES rusty-brown; ellipsoid; roughened; 8-9 x 5-5.5 μ. POISONOUS.

C. traganus is solitary to gregarious in conifer forests in the Pacific Northwest and Rocky Mountains in the fall. It is poisonous, but its odor and taste would probably discourage the mushroom hunter anyhow.

Cortinarius subfoetidus A.H. Smith   Plate 48

PILEUS 4-10 cm in diameter; convex to plane or umbonate; glutinous; fibrillose; bluish-lavender to lilac, fading to buff or pallid on disc. FLESH thick; concolorous with pileus, becoming pallid; odor sweet and nauseous. GILLS adnexed to slightly decurrent; close; concolorous with pileus, becoming brown. STIPE 5-8 cm long, 1-2 cm thick; solid; equal; concolorous with pileus, becoming pallid; lavender fibrillose sheath which leaves a median zone. SPORES rusty-brown; subellipsoid; roughened; 7-9 x 5-5.5 μ. POISONOUS.

C. subfoetidus occurs in the Pacific Northwest under conifers in the fall. The sickening odor of this beautiful species is obviously responsible for its name.

Another western species, **C. mutabilis** A.H. Smith, is a duller violet-gray; the stipe and flesh are concolorous, and stain purplish when bruised. Its odor and taste are not distinctive; it also is poisonous.

Cortinarius cinnabarinus Fr.

PILEUS 3-6 cm in diameter; convex to plane or umbonate; dry; fibrillose; bright red. FLESH firm; pale reddish. GILLS adnate to adnexed; rather far apart; bright red, becoming dark rusty-red. STIPE 2-6 cm long, 4-8 mm thick; stuffed, then hollow; equal or slightly enlarged downward; fibrillose;

bright red. SPORES rusty-brown; elliptical; slightly roughened; 7-9 x 4.5-5.5 μ. POISONOUS.

C. cinnabarinus is one of the most brilliantly colored species in the genus. It occurs in the fall in western states under conifers or hardwoods; it seems to be particularly associated with oaks.

Cortinarius californicus A.H. Smith   Plate 49

PILEUS 3-8.5 cm in diameter; conic to broadly umbonate; smooth; hygrophanous; rusty-red. Flesh thick; watery; concolorous with pileus. GILLS adnate to adnexed; rather far apart; orange-red; edges slightly serrate. STIPE 8-15 cm long, 5-15 mm thick; hollow; equal; dull orange; cortina orange. SPORES rusty-brown; ellipsoid to subovoid; roughened; 7-9 x 4-4.5 μ. POISONOUS.

C. californicus occurs in the fall under conifers in California, Oregon, and Idaho.

Cortinarius phoeniceus var. occidentalis
(Bull. ex Maire) A.H. Smith   Plate 50

PILEUS 3-8 cm in diameter; convex to campanulate or slightly umbonate; dry; fibrillose; bright red. FLESH thin; firm; reddish near cuticle, olive-brown near the gills and stipe. GILLS adnate; rather far apart; concolorous with pileus or brighter. STIPE 4-7 cm long, 6-12 mm thick; fibrillose; equal; yellow; mycelium yellow. SPORES rusty-brown; oblong to subellipsoid; minutely roughened; 6-6.5 x 4-4.2 μ. POISONOUS.

C. phoeniceus var. *occidentalis* is more intensely colored in all parts than the **C. phoeniceus** of the eastern states and Europe. It grows gregariously in conifer forests along the Pacific Coast in the fall.

There are several species containing the same deep red pigment. **C. sanguineus** (Wulfen ex Fr.) S.F. Gray (Plate 51) is blood-red in all parts; **C. semisanguineus** (Fr.) Gill. has red gills and a yellow pileus and stipe. **C. cinnamomeus** (Fr.) S.F. Gray also exudes red pigment when the pileus is crushed in KOH, although the fungus is not red in any part. Instead, the pileus is yellowish-brown to olive-brown, the gills are bright yellow at first, and the stipe is yellowish above, and concolorous with the pileus (or a bit lighter) below. All are poisonous.

Cortinarius collinitus Fr.

PILEUS 3-9 cm in diameter; convex to plane; gelatinous; orange-brown, darker on disc. FLESH watery; yellowish-buff. GILLS adnexed; close; pallid or pale violet. STIPE 6-12 cm long, 1-1.5 cm thick; solid; equal; violet; lower two-thirds with glutinous bands from universal veil; cortina thick, white. SPORES rusty-brown; almond-shaped; roughened; 12-15 x 7-8 μ. POISONOUS.

C. collinitus is common on the West Coast and fruits in both coniferous and deciduous woods in the fall. It has several varieties.

C. collinitus var. *trivialis* (Lange) A.H. Smith has pallid gills, a yellow-brown stipe with paler glutinous bands, and smaller spores which measure 10-13 x 6-7 μ.

D. SPORES BLACK TO SMOKY GRAY (or Deep Purplish-Brown)

There are only two families in this spore-color category: Coprinaceae and Gomphidiaceae. The Coprinaceae is the largest and most diversified, with spores ranging from black to deep purple-brown—darker than in the Strophariaceae (see above). In the genus *Coprinus* the gills deliquesce, turning into a black liquid by a process of autodigestion. Some members of the Coprinaceae are edible, others may be poisonous or hallucinogenic.

The Gomphidiaceae have black or very dark smoky-gray spores, and decurrent gills that are often waxy in texture.

All have mycorrhizal relationships with conifer roots and are regarded as edible, although not choice.

FAMILY COPRINACEAE

Members of the Coprinaceae are a rather diversified group of black-spored or dark purplish-brown-spored agarics. The gills may or may not *deliquesce* (turn to an inky fluid).

Genus *Psathyrella*

Psathyrella contains a very large number of species, because many of its members were formerly classified in the genera *Hypholoma*, *Psilocybe*, and *Stropharia* in the family Strophariaceae (see above), whose members are sometimes toxic or hallucinogenic. It has purplish-brown spores, whereas the other three genera in the Coprinaceae have black spores, and its attached gills will distinguish it from the genus *Agaricus* (see above), which has the same spore color but free gills. There are few known edible species in *Psathyrella*; only one, *P. candolleana*, has been highly rated as an edible.

Psathyrella candolleana (Fr.) Maire

PILEUS 3-8 cm in diameter; convex to plane with a blunt umbo; hygrophanous; pale yellow to honey-color or light brown; scattered white scales near the margin, which is striate, with noticeable white fragments of veil. FLESH thin; fragile; white. GILLS adnate; close; whitish, becoming purplish-brown. STIPE 2-5 cm long, 2-4 mm thick; hollow; smooth; white; annulus white, fibrillose, usually evanescent. SPORES purplish-brown; ellipsoid; smooth; 7-8 x 4-4.5 μ. EDIBLE and good; see below.

P. candolleana is widely distributed throughout the West in grassy places, and can often be found in abundance around hardwood stumps. Its flesh is fragile and thin, but when a sufficient quantity can be found it makes a very good, delicately flavored dish.

Psathyrella hydrophila (Fr.) A.H. Smith is also common in the West and can often be found in dense caespitose clumps on decayed hardwood logs or stumps. It has a light to dark brown, hygrophanous pileus and a white stipe. **P. longistriata** (Murr.) A. H. Smith may be solitary or gregarious under hardwood or conifers in the coastal states. It has a conspicuous white annulus with striations on the upper surface; grayish-brown to dark reddish-brown pileus; and long white stipe. The edibility of both species is unknown.

Genus *Panaeolina*

Panaeolina contains a single species, *P. foenisecii*, which was originally classified in the genus *Psilocybe* (see above) and later transferred to the genus *Panaeolus* (see below). However, all the species now classified in *Panaeolus* have smooth, black spores, while those of *P. foenisecii* are warty and a deep purplish-brown.

Panaeolina foenisecii (Pers. ex Fr.) Maire   Plate 52
Psilocybe foenisecii Quél.
Panaeolus foenisecii (Fr.) Kühn.

PILEUS 1-3 cm in diameter; conic to convex, becoming umbonate in age; smooth or cracked when dry; hygrophanous; reddish-brown when moist, fading to tan or grayish-brown; margin striate when moist. FLESH thin, watery, tan. GILLS adnate; broad; pale chocolate-brown, becoming spotted by darker spores; edges pallid. STIPE 4-8 cm long, 2-3.5 mm thick; cartilaginous; equal; pale brownish-gray; slightly hairy. SPORES dark purplish-brown; elliptical; warted; 12-15 x 6-7 μ. HALLUCINOGENIC.

P. foenisecii is widespread and may be found mostly on

lawns from spring to fall in the West. Where the climate is mild, it may occur the year around. The hygrophanous pileus causes it to vary greatly in color, and as it dries it may become very pale and cracked. This very common little mushroom is frequently eaten by children who find it where they play, and this can be of concern because it contains small amounts of the psychotropic substances psilocybin and psilocin.

Genus *Panaeolus*

Members of *Panaeolus* can be readily recognized by their mottled gills, due to irregular distribution of the spores when they mature; the gills are attached and do not deliquesce, and the smooth black spores have an apical pore. With the exception of *P. semiovatus*, which is annulate, these fungi lack an annulus and volva. They are mostly dung-inhabiting and can be found in just about every pasture where horses or cows have been. They should never be eaten; several species contain hallucinogenic substances.

KEY TO SPECIES OF *PANAEOLUS*

1a. Fruiting body annulate .... *P. semiovatus*, p. 172
 b. Fruiting body not annulate 2
   2a. Pileus large, up to 20
       cm broad .... *P. phalaenarum*, p. 172
    b. Pileus usually not more than 5 cm broad 3
3a. Pileus smooth 4
 b. Pileus wrinkled or reticulate over
    disc...........................*P. retirugis*, p. 173
   4a. Pileus olive-brown to grayish-brown;
       margin with hanging remnants
       of veil.............. *P. campanulatus*, p. 173
    b. Not as above 5

5a. Pileus conic when young;
    dark brown .... *P. acuminatus*, p. 173
  b. Pileus convex or slightly umbonate;
    light yellowish-brown, with darker zone
    near margin .... *P. subbalteatus*, p. 174

Panaeolus semiovatus (Sow. ex Fr.) Lund. and Nannf.
Panaeolus separatus (L. ex Fr.) Quél.

PILEUS 3-6 cm in diameter; broadly conic to campanulate; smooth or cracked in age; viscid; whitish, becoming pale buff. FLESH thin; soft; white. GILLS adnate; close; broad; pale grayish or brown, becoming mottled. STIPE 10-16 cm long, 6-10 mm thick; equal or slightly enlarged at base; pallid to buff. ANNULUS median; whitish, often stained by spores. SPORES black; elliptical; smooth; with apical pore; 17-22 x 8-11 μ. EDIBILITY unknown.

P. semiovatus is solitary to gregarious on horse dung in western North America. Psilocybin has been found in some specimens in Colorado.

Panaeolus phalaenarum (Fr.) Quél.

PILEUS 4-10 cm in diameter; hemispheric to convex; at first white, becoming yellowish and cracked. FLESH thick; firm; white. GILLS adnate; white, becoming mottled. STIPE 8-20 cm long, 5-15 mm thick; equal; straight or twisted; solid; white. SPORES black; elliptical; smooth; with apical pore; 14-20 x 8-11 μ. SUSPECT; see below.

P. phalaenarum is distinctive for its large size, white pileus, and solid stipe. It is solitary or gregarious on dung, and has been recorded in Alaska, Washington, and California. It is suspected of containing psychotropic substances.

Panaeolus campanulatus (Fr.) Quél.   Plate 53

PILEUS up to 5 cm in diameter; conic to subovoid when young, becoming campanulate in age; smooth; viscid or dry; olive-gray or grayish-brown; margin with conspicuous white veil remnants. FLESH thin except on disc; soft; grayish or white. GILLS adnate; pale gray, becoming spotted as spores mature; edges white. STIPE 6-14 cm long, 2-6 mm thick; hollow; grayish-brown with powdery bloom when young. SPORES black; broadly elliptical; smooth; apical pore; 13-18 x 8-10 μ. HALLUCINOGENIC.

P. campanulatus is extremely common in pastures or wherever there is horse or cow dung in all western states and provinces. It sometimes also appears in beds of cultivated mushrooms. The scalloped white fragments of veil tissue on the margin of the pileus are often quite striking, particularly in young specimens.

P. retirugis (Fr.) Gill. has a paler pileus, with wrinkles or reticulations on the disc. Both species contain psilocybin and some other unidentified substances and are hallucinogenic.

Panaeolus acuminatus (Schaeff.) Fr.

PILEUS 1.5-3 cm in diameter; conic when young, expanding in age; smooth; hygrophanous; dark brown, becoming tan upon drying; margin striate, almost transparent. FLESH firm; brown. GILLS adnexed; crowded; pallid; edges white. STIPE 2.5-7.5 cm long, 2-4 mm thick; equal; pallid at apex, becoming brown in lower part; base enlarged; white mycelium. SPORES black; elliptical; smooth; apical pore; 13-16 x 8-11 μ. HALLUCINOGENIC.

P. acuminatus occurs on dung from northern California to northern Washington. This common little mushroom may be solitary or gregarious.

Panaeolus subbalteatus (Berk. and Br.) Sacc.

PILEUS 2-5 cm in diameter; convex to plane or umbonate; hygrophanous; light yellow-brown, with darker zone near margin; margin striate, at first inrolled. FLESH thick; yellowish. GILLS adnate or adnexed; brownish becoming mottled; edges white and slightly fringed. STIPE 2-8 cm long, 2-10 mm thick; equal; mostly smooth, striate at apex; hollow; cartilaginous; brown; white mycelium at base. SPORES black; subelliptical; smooth; apical pore; 10-13 x 6.5 μ.

P. subbalteatus occurs on dung, sometimes in beds of cultivated mushrooms, from California to Washington and Idaho. It is common from spring to fall and is recognized by the distinctive darker zone on the margin.

A somewhat smaller species, **P. fimicola** (Fr.) Quél., also hallucinogenic, is known only in California. It also has a brown zone near the margin, but the pileus is brownish-gray and the margin is scalloped. It also differs in its solitary habit.

Genus *Coprinus*

Members of *Coprinus* have smooth black spores and extremely crowded, parallel gills; they may occur on dung, humus, or wood. They are commonly called inky caps because the gills deliquesce as the spores mature: this is a process of autodigestion brought about by enzymatic action that causes the tissues to dissolve into a black fluid from the margin upward. It was formerly believed that all the spores were in the "ink," but it is now known that the majority of them are dispersed in the air like those of other agarics. As the spores are discharged, the gills deliquesce upward and gradually free the rest. Finally only the skeletal top of the pileus remains.

There are a great many very small and ephemeral species of *Coprinus* which are not well known but some of the larger ones are easily recognized and considered choice edibles. One species, *C. atramentarius*, can cause a strange flushing and uncomfortable feeling of heat if it is consumed with alcohol, but is otherwise safe to eat.

KEY TO SPECIES OF *COPRINUS*

1a. Pileus small, not over 2.5 cm broad;
  plicate ....*C. plicatilis*, p. 175
 b. Pileus larger, 3-10 cm in diameter 2
  2a. Pileus covered with shiny
   particles ....*C. micaceus*, p. 176
  b. Not as above 3
3a. Pileus with cuticle of shaggy scales 4
 b. Not as above 5
  4a. Pileus 7-12 cm high; spores 13-18
   x 7-8 μ. ....*C. comatus*, p. 176
  b. Pileus 5-8 cm high; spores 18-25
   x 12-15 μ. ....*C. sterquilinus*, p. 177
5a. Pileus egg-shaped; grayish-brown
  on disc ....*C. atramentarius*, p. 177
 b. Pileus conic at first, later with
  raised margin 6
  6a. Pileus silvery-white, with
   downy covering ....*C. lagopus*, p. 178
  b. Pileus grayish, with brown
   tinged disc ....*C. fimetarius*, p. 178

Coprinus plicatilis (Curt. ex Fr.) Fr.

PILEUS 1-2.5 cm in diameter; conic to campanulate, becoming plane to depressed; sulcate up to disc; brown, becoming gray or bluish-gray; disc brown. FLESH thin; fragile. GILLS free, but attached to a collar; far apart; narrow; gray. STIPE 5-8 cm long, 1-2 mm thick; equal; hollow; white or yellowish. SPORES black; broadly elliptical; smooth; with apical pore; 10-12 x 7.5-9.5 μ. INEDIBLE.

C. plicatilis is widely distributed and often occurs in abundance in grassy spots in wet weather. These fragile little fungi, commonly called Japanese Parasols, are unpalatable.

Coprinus micaceus (Bull. ex Fr.) Fr.   Plate 54

PILEUS 3-6 cm in diameter; conic to campanulate; striate on edges; at first covered with shining white particles; pale yellowish-brown or reddish-brown. FLESH soft; thin; pallid. GILLS adnexed; broad; crowded; white or gray, becoming black. STIPE 3-5 cm long, 2-5 mm thick; equal or slightly enlarged at base; hollow; white. SPORES black; elliptical; smooth; with apical pore; 7-10 x 3.5-5 μ. EDIBLE and choice; see below.

C. micaceus is widely distributed, occurring in dense clumps around stumps or decayed wood. The rain often washes away the glistening particles, but it is still recognizable by its caespitose habit. To many, *C. micaceus* has the best flavor of any *Coprinus*.

Coprinus comatus (Müll. ex Fr.) S.F. Gray

PILEUS 5-10 cm high, 2.5-5 cm in diameter; cylindric, expanding to bluntly conic or campanulate; dry; white with large shaggy yellowish-brown or reddish-brown tips. FLESH soft; thin; white. GILLS adnexed or nearly free; crowded; white, often pinkish-tinged, becoming black. STIPE 8-20 cm long, 1-1.5 cm thick; smooth; dry; equal or tapering upward; white; annulus is free and moves downward. SPORES black; elliptical; smooth; with apical pore; 13-18 x 7-8 μ. EDIBLE and choice; see below.

C. comatus (Fig. 21) is a well-known edible, commonly called Shaggy Mane, which occurs in many parts of the West in spring and fall. It grows on the ground, often in hardpacked soil such as football fields and tennis courts, or along roadsides. It is one of the commonest species of mushroom along the Alcan Highway in Alaska and the

FIG. 21. *Coprinus comatus.*

Yukon Territory. It may be scattered, but is often in large groups. It should be cooked as soon as possible, because it can deliquesce very quickly.

A very similar and also edible species, **C. sterquilinus** Fr., is smaller but has much larger spores, 18-25 x 12-15 μ. It is usually found growing on manure.

Coprinus atramentarius (Bull. ex Fr.) Fr.   Plate 55

PILEUS 2-8 cm in diameter; ovate to campanulate in age; fibrillose; dry; shiny; gray to grayish-brown; margin slightly striate. FLESH thin; soft; pallid. GILLS free or nearly so; crowded; white to grayish, becoming black. STIPE 8-15 cm long, 8-12 mm thick; equal; hollow; white; annulus fibrous and inferior. SPORES black; elliptical; smooth; with apical pore; 8-12 x 4.5-6.5 μ. EDIBLE, but see below.

C. atramentarius usually occurs in caespitose clusters, although it may be gregarious, in rich soil or around stumps. It is found from Alaska to Mexico and from sea level to over 7,000 feet.

C. atramentarius is considered inferior in flavor to *C. micaceus* and *C. comatus*. It may produce a strange vaso-motor reaction when consumed with alcohol, and symptoms may recur for two or three days afterward whenever any alcohol is drunk, even though no more mushrooms have been eaten. The substance responsible for this phenomenon has just recently been isolated and named *coprine*, after the generic name of the mushroom.

Coprinus lagopus Fr.

PILEUS 2.5-4 cm in diameter; conic to campanulate, becoming expanded with raised margin; silvery-gray, with downy white covering; disc livid. FLESH thin; gray. GILLS free; far apart; white, becoming black. STIPE 5-15 cm long, 4-5 mm thick; hollow; scurfy; gray or whitish. SPORES black; elliptical; smooth; with apical pore; 12-13 x 7-8 μ. EDIBILITY unknown.

C. lagopus grows on the ground; the superficial downy remnants of a universal veil on the pileus, and the long slender stipe, make it distinctive.

A similar dung-inhabiting species in the West, **C. fimetarius** Fr., is grayish with a brownish disc and is at first covered with white scales. Its stipe is equally long and slender as that of *C. lagopus*, but it is much more ephemeral: it may occur in troops on dung heaps in late afternoon and be completely deliquesced by the next morning. Its edibility is unknown.

Genus *Pseudocoprinus*

Pseudocoprinus contains a few species which closely resemble *Coprinus* (see above) except that their gills do not deliquesce. The pileus is plicate; spores are black and smooth, with an apical pore. These delicate fungi often fruit prolifically on logs or buried wood.

Pseudocoprinus disseminatus (Fr.) Kühn.

PILEUS 5-10 mm in diameter; conic to convex or campanulate; moist; plicate or striate on the edges; whitish, becoming gray or grayish-brown; disc dark yellowish or yellowish-brown. FLESH thin; fragile. GILLS adnate; broad; rather far apart; white to gray, becoming black. STIPE 2-3 cm long, 0.5-1 mm thick; hollow; minutely hairy; white. SPORES black; elliptical; smooth; with apical pore; 7-10 x 4-4.5 μ. EDIBLE but poor; see below.

P. disseminatus is common throughout the temperate zone and sometimes occurs in clusters of hundreds of fruiting bodies. Although reportedly edible, it is of no importance for the table because of the lack of substance of its fragile flesh.

FAMILY GOMPHIDIACEAE

Members of the Gomphidiaceae have black or very dark smoky-gray spores and thick, widely spaced, and decidedly decurrent gills. The two genera *Gomphidius* and *Chroogomphus* are both represented by some common species in the West; all have a mycorrhizal association with the roots of conifers. *Gomphidius* has a gelatinous cuticle and white, nonamyloid flesh in the pileus; *Chroogomphus* is dry to viscid and has colored, amyloid flesh.

Genus *Gomphidius*

Members of *Gomphidius* have smoky-gray to black spores, waxy and decurrent gills, a slimy or gelatinous pileus, and white flesh which may turn pink or vinaceous when cut. The stipe often has the remains of a glutinous annulus and is frequently yellow at the base.

Gomphidius oregonensis Pk.

PILEUS up to 10 cm in diameter; convex with inrolled margin when young, usually becoming irregularly plane or centrally depressed with age; surface glutinous; dingy pinkish-gray to grayish-lilac. FLESH white, darkening slowly with age or when cut; odor and taste not distinctive. GILLS decurrent; somewhat far apart; thick; waxy; white, becoming grayish with age. STIPE up to 8 cm long; moderately thick above, tapering at base; whitish above, yellow at base, outer surface soon darkening; glutinous veil, leaving a superior ring. SPORES sooty-brown; subfusiform; smooth; 10-13 x 5-8 μ. EDIBLE; see below.

G. oregonensis is a common species in fall and winter along the Pacific Coast, and in spring and fall in the Rocky Mountains. It is often found under Douglas Fir.

G. glutinosus, also found in the coniferous forests of the West, can be distinguished from *G. oregonensis* by its larger spores, which measure 16-20 x 4-7.5 μ. Both are edible, but the slimy pellicle must be removed from the cap.

Gomphidius subroseus Kuaf.   Plate 56

PILEUS up to 8 cm in diameter; convex with margin inrolled at first, becoming plane or centrally depressed; surface glutinous; rosy-pink, becoming whitish along margin. FLESH white, darkening slowly with age or when cut; odor and taste not distinctive. GILLS decurrent; far apart; thick; waxy; white, becoming grayish with age. STIPE up to 8 cm in length; tapering toward base; glutinous; whitish above, yellow toward base, darkening with age; superior ring left by glutinous veil. SPORES sooty-brown; cylindric to fusiform; smooth; 16-21 x 4.5-7 μ. EDIBLE and choice; see below.

G. subroseus is common in Douglas Fir forests in the rainy season in the Rocky Mountains and along the Pacific Coast. It occurs from sea level to over 6,000 feet elevation. Despite its slimy covering, *G. subroseus* is a beautifully colored mushroom with its rosy pileus and a stipe that varies from white above to bright lemon-yellow near the base. It is edible and considered choice, but the pellicle must be removed.

Gomphidius maculatus (Scop. ex Fr.) Fr.

PILEUS up to 10 cm in diameter; broadly convex with inrolled margin when young, becoming plane; surface glutinous; brown to reddish-brown. FLESH white, with a vinaceous tinge just under surface of pileus; firm; lacks distinctive odor or taste. GILLS decurrent; far apart; white, becoming smoky-gray with age. STIPE up to 8 cm long; tapering toward base; white above, yellowish toward base; black fibrils on lower part; lacks glutinous ring. SPORES black; fusiform; smooth; 14-22 x 6-8 μ. EDIBLE.

G. maculatus is quite widespread but most commonly associated with the larches. Its reddish-brown color and the absence of a glutinous veil distinguishes it from other member of *Gomphidius*.

Genus *Chroogomphus*

Members of *Chroogomphus* are characterized by black spores, a dry or viscid pileus that ranges from orange to reddish-brown, and widely spaced, decurrent gills. All species are restricted to the proximity of conifers with which they have a mycorrhizal association. All are regarded as edible.

The absence of a glutinous surface to the pileus, and flesh which turns blue in Melzer's reagent, distinguish members of *Chroogomphus* from those of the closely related genus *Gomphidius* (see above).

Chroogomphus rutilus (Schaeff. ex Fr.) O.K. Miller
Plate 57

PILEUS up to 12 cm in diameter; convex, often with an umbo; margin inrolled when young; surface viscid; reddish-brown. FLESH orange; firm; lacks distinct odor or taste. GILLS decurrent; far apart; yellowish-orange to dark brown. STIPE up to 15 cm long; tapering toward base; faint remnant of an annular zone when young; orange-yellow with reddish tinge. SPORES dark gray to black; elliptical; smooth; 14-22 x 6-7.5 μ. EDIBLE.

C. rutilus, found principally under pines, is the most common species of *Chroogomphus* along the Pacific Coast in the fall and winter. It also occurs in summer and fall in the Rocky Mountains.

Two related species are **C. vinicolor** (Pk.) O.K. Miller, which has a dark reddish-brown, viscid, pointed pileus, and **C. ochraceous** (Kauf.) O.K. Miller, which has an orange pileus and stipe. Both are edible but unremarkable.

Chroogomphus tomentosus (Murr.) O.K. Miller

PILEUS up to 6 cm in diameter; convex; surface dry; yellowish-orange. FLESH light orange; firm; odor and taste not distinctive. GILLS decurrent; thick; far apart; yellowish-orange. STIPE up to 15 cm; tapering toward base; dry; orange-yellow. SPORES blackish; elliptical; smooth; 15-25 x 6-9 μ. EDIBLE.

C. tomentosus, although not as abundant as *C. rutilus*, is widespread throughout the coniferous forests of the West. It is most commonly found in association with Douglas Fir and Western Hemlock.

Another edible species with a dry cap, **C. leptocystis** (Sing.) O.K. Miller, also occurs in the West, but it is reddish-brown and has thin-walled cystidia.

E. SPORES
WHITE, YELLOWISH, OR PINKISH-PUFF,
(Green, in one genus)

There are five families in this very large category, which comprises most of the agarics, and the great majority of genera have decidedly white spores. However, in some species of the Russulaceae the spores are various shades of yellow, and in the Tricholomataceae some are pinkish-buff. The genus *Chlorophyllum* in the Lepiotaceae is unique because it has green spores.

Distinguishing characters are: very brittle flesh, with presence or lack of latex, in the Russulaceae; the presence of a volva in the Amanitaceae; waxy gills in the Hygrophoraceae; a pileus easily separable from the stipe in the Lepiotaceae; and a confluent pileus and stipe in the Tricholomataceae.

FAMILY RUSSULACEAE

The Russulaceae are usually medium to large fungi with a characteristic brittle flesh, due to the presence of large spherical cells called *sphaerocysts* in the flesh of the pileus. The gills may be free or attached, there is a central stipe, and both annulus and volva are lacking. Spores may be white, buff, or some shade of yellow, and they are ornamented (warted or ridged) and amyloid. These mushrooms usually grow on the ground, but a few occur on wood, and some are associated with particular conifers or hardwoods.

The family Russulaceae contains only two genera, *Russula* and *Lactarius*. The two genera, differentiated by the presence of a fluid called *latex* in *Lactarius*, are easily recognizable as such, but a great many species, particularly in *Russula*, are very difficult even for experts to identify.

Genus *Russula*

KEY TO SPECIES OF *RUSSULA*

1a. Spore print some shade of yellow 2
 b. Spore print white or slightly creamy 3
  2a. Pileus viscid when wet, velvety
    when dry; greenish-gray to dark
    green .... *R. aeruginea*, p. 184
   b. Pileus reddish-purple to
    olivaceous .... *R. xerampelina*, p. 188
3a. Flesh turns red or black when bruised 4
 b. Flesh unchanging 5
  4a. Flesh blackens .... *R. albonigra*, p. 185
   b. Flesh turns reddish, then slowly changes
    to black .... *R. densifolia*, p. 186
5a. Pileus viscid; dingy yellow; margin
  deeply grooved; strong unpleasant odor
  in age .... *R. foetens*, p. 187
 b. Not as above 6
  6a. Pileus and stipe white; large, up
    to 20 cm broad .... *R. brevipes*, p. 185
   b. Pileus bright red; stipe rosy-pink
    to reddish .... *R. sanguinea*, p. 187

Russula aeruginea Lindbl.

PILEUS 3-7 cm in diameter; convex when young, becoming expanded and centrally depressed; margin often slightly striate; surface viscid when wet, velvety when dry; dark green or greenish-gray. FLESH thick; white to greenish; lacks distinct odor or taste. GILLS adnate to nearly free; white to cream-color. STIPE 3-5 cm in length; moderately thick;

smooth; white; firm. SPORES pale yellow; subglobose; lightly ornamented with small warts connected by lines; 7-9 x 5.5-7 μ. EDIBLE.

R. aeruginea is widespread in coniferous forests and mixed woodlands from Alaska south. Its greenish color, mild taste, and yellow spores are identifying characters.

An extremely large related species, **R. olivacea** (Schaeff. ex Secr.) Fr., may attain a diameter of 35 cm. It has an olive-colored pileus, sometimes mixed with reddish tinges, and a thick, pinkish stipe; the gills are a dingy yellow. *R. olivacea* occurs under conifers in Washington and Idaho, and is reported to be edible.

Russula albonigra (Krombh.) Fr.

PILEUS 7-15 cm in diameter; broadly convex with inrolled margin when young, becoming plane and centrally depressed; margin smooth; surface dry except when young; white at first, soon becoming brown or black where bruised. FLESH thick; firm; white, becoming gray-brown to black when cut or bruised; no distinctive odor or taste. GILLS adnate to decurrent; close; white, becoming blackish when bruised. STIPE 3-8 cm in length; proportionately thick; equal or enlarged basally; smooth; white, rapidly staining black where touched or cut. SPORES white; ellipsoid; slightly ornamented; 7.5-10 x 6.5-8 μ. EDIBILITY unknown.

R. albonigra occurs in deciduous or mixed forests, most commonly in the fall. It can be recognized by its relatively dry pileus and very firm flesh which rapidly changes to black when handled or bruised.

Russula brevipes Pk.

PILEUS 8-20 cm in diameter; margin inrolled, center depressed; minutely felted; dry; white to buffy, staining yellowish-brown with age. FLESH thick; firm; white; odor slight;

taste slightly acrid. GILLS decurrent; thin; close; white, often with blue-green tinge near stipe; droplets of liquid sometimes on gills when young; becoming stained with age. STIPE up to 8 cm long, 3-5 cm thick; equal or tapering toward base; solid; dull white. SPORES white to light cream; ellipsoid to nearly globose; ornamented with coarse warts or reticulated; 8-10.5 x 6.5-8.5 μ. EDIBLE; see below.

R. brevipes (mistakenly called *R. delica* by many authors) is a very common and widespread species in the forests of the West. In mountainous regions it occurs in both spring and autumn. Although it is large and white with a centrally depressed pileus, it is often hidden from view as it grows under masses of conifer needles or leaves of broadleafed trees. By learning to examine mounds rising on the forest floor, one often finds these mushrooms to be abundant. An identifying feature is the blue-green tinge to the gills near the stipe when the fruiting bodies are fresh. Some use this species for the table. It is quite edible but has little flavor, and is usually pickled because of its crispness.

A closely related species, **R. cascadensis**, occurs in the Pacific Northwest. It lacks the blue-green tinge to the gills and has a strongly acrid taste, which would make it inedible.

Russula densifolia (Secr.) Gill.   Plate 58

PILEUS 5-15 cm in diameter; convex, depressed centrally when young, becoming almost funnel-shaped; margin smooth; viscid when moist; dull white, becoming brown with age. FLESH thick; firm; white, turning reddish, then black when cut or bruised; lacks distinct odor; often has acrid taste. GILLS adnate to decurrent; close; white, turning red, then black, when bruised. STIPE 3-10 cm in diameter; fairly thick; equal or enlarged at base; smooth; white, bruising red, then black. SPORES white; ellipsoid; finely ornamented with warts joined by lines; 7.5-10 x 6-8 μ. POISONOUS.

R. densifolia, may be either solitary or occur in small groups in deciduous or mixed woods in autumn.

Russula foetens Fr.

PILEUS 5-12 cm in diameter; convex when young, soon becoming flat with center depressed; margin striate; viscid; dirty yellow. FLESH thin; firm, becoming fragile with age; whitish; odor of bitter almond at first, then fetid; taste acrid. GILLS adnexed; rather close; white to yellowish; exuding drops of liquid when young. STIPE 3-10 cm long; moderately thick; creamy-white, often stained brown when mature. SPORES white; ellipsoid to subglobose; coarsely ornamented; 8.5-10 x 8-9 μ. INEDIBLE; see below.

R. foetens is a common species in oak woodlands and coniferous forests. The dingy-yellow pileus, as well as its striate margin and the fetid odor that develops with age, distinguish it from other species of *Russula*. When young, the gills, like those of *R. brevipes* (see above), often have droplets of liquid on them. Although said to be nonpoisonous, it has a very unpleasant flavor.

Russula sanguinea Fr.   Plate 59
Russula rosacea Fr.

PILEUS 5-9 cm in diameter; broadly convex with center flat, sometimes depressed; margin smooth to slightly striate; viscid when wet; uniformly bright, shiny red. FLESH fragile; white, tending to reddish when cut; odor mild; taste intensely acrid. GILLS adnate to subdecurrent; close to more widely spaced; cream-color. STIPE 3-5 cm in length; equal; smooth; rosy pink to reddish. SPORES very pale cream (almost white); subglobose; ornamented with numerous spiny warts; 7-8.5 x 6-7 μ. EDIBILITY unknown.

R. sanguinea is invariably associated with conifers, particularly pines. Along the Pacific Coast this small, brilliantly red species often grows in association with Beach Pine, especially where there is a short, dense turf. The pinkish-red stipe and peppery taste of the flesh are additional characters that help identify it.

A closely related species, **R. americana** Sing., described from the Olympic Peninsula of Washington, has larger spores and more fragile flesh. **R. emetica** (Fr.) Pers., another red species, has white gills and stipe; it is poisonous.

Russula xerampelina Fr.   Plate 60

PILEUS 5-10 cm in diameter; convex, becoming plane or slightly depressed centrally; margin usually smooth; reddish-purple to olivaceous. FLESH firm; whitish; stains brown; taste not distinctive; odor of decaying fish when old or drying. GILLS adnexed; close to moderately distant; white when young, soon turning cream. STIPE 3-8 cm long; relatively thick, equal; white to reddish. SPORES creamy yellow; ellipsoid to subglobose; ornamented with small spines; 8-10 x 6-8.5 μ. EDIBLE; see below.

R. xerampelina occurs in coniferous or mixed forests, where many other members of this genus grow. It is said to be edible and to have a good flavor when young.

Genus *Lactarius*

A fluid, called *latex*, exudes when specimens of *Lactarius* are cut or bruised. In older or very dry specimens it may not be evident, but the young fruiting bodies usually have a copious flow, particularly if the gills are cut near the stipe. This fluid may be watery, white, or colored, and there may be a change of color upon exposure to air in some species. The taste of the flesh should also be noted—it may range from mild to peppery or acrid. The spores are white, buff, or yellow, always ornamented, and amyloid.

KEY TO SPECIES OF *LACTARIUS*

A. Latex watery or white, not changing color
  1a. Fruiting body with strong aromatic odor, particularly when dried; pileus dark brownish-red .... *L. camphoratus*, p. 190
    b. Without distinctive odor 2
      2a. Pileus flesh-colored; densely hairy around margin .... *L. torminosus*, p. 190
        b. Pileus white; taste peppery .... *L. piperatus*, p. 191
        c. Not as above 3
    3a. Pileus reddish-brown; dry or moist........................ *L. rufus*, p. 191
      b. Pileus orange; viscid ...... *L. aurantiacus*, p. 192
      c. Pileus olive-brown with darker disc; gills bruising brown or blackish-brown .... *L. necator*, p. 192
B. Latex white, changing color or staining tissue upon exposure
  1a. Latex becoming yellow..................... 2
    b. Not as above 3
      2a. Pileus buff to cinnamon-buff; stipe smooth .... *L. chrysorheus*, p. 192
        b. Pileus pale yellow to dark ochraceous; stipe with depressed brighter spots .... *L. scrobiculatus*, p. 193
    3a. Latex becoming lilac or purplish .... *L. uvidus*, p. 194
      b. Latex drying greenish-gray or bluish .... *L. mucidus*, p. 194
C. Latex orange or red; all parts staining green in time
  1a. Latex orange.............. *L. deliciosus*, p. 195
    b. Latex red............... *L. sanguifluus*, p. 194

Lactarius camphoratus (Bull. ex Fr.) Fr.

PILEUS 1-4 cm in diameter; convex to expanded; often umbonate, sometimes depressed; smooth; dry; azonate; dark reddish-brown. FLESH thin, firm; paler than pileus or concolorous; odor aromatic. LATEX white, unchanging. GILLS adnate to subdecurrent; narrow; close; whitish to reddish-brown in age. STIPE 1-5 cm long, 4-8 mm thick; dry; smooth; dark reddish-brown. SPORES white; subglobose; coarsely warted; amyloid; 6-8.5 x 6-7.5 μ. EDIBLE; see below.

L. camphoratus is a common and widely distributed species with an aromatic odor that has been variously described as being like sweet clover or maple syrup. It is said to be edible. It grows on rotten conifer logs or on the ground.

A somewhat similar species, **L. helvus** Fr., is larger and paler, tan or reddish-gray, and has a watery white latex. Its odor is reported to be like camphor, and its edibility is unknown.

Lactarius torminosus (Schaeff. ex Fr.) S.F. Gray

PILEUS 4-10 cm in diameter; convex to depressed; viscid; smooth on disc and densely hairy at margin; margin persistently inrolled; zoned with yellowish, buff, and pinkish bands. FLESH soft; thick; white or pinkish; taste acrid. LATEX white. GILLS decurrent; close; narrow; white to yellowish or pinkish. STIPE 3-6 cm long, 1.5-2 cm thick; equal or tapered at base; smooth or slightly hairy; stuffed, becoming hollow in age; yellowish or pinkish. SPORES white; elliptical; warted and reticulate; amyloid; 8-10 x 6-8 μ. POISONOUS.

L. torminosus occurs in the fall in mixed forests and seems to be especially associated with birches and hemlocks. Its inrolled and very hairy margin is distinctive. It is poisonous, although some people seem to be able to eat it.

Lactarius piperatus (L. ex Fr.) S.F. Gray

PILEUS 4-12 cm in diameter; convex to depressed or funnel-shaped; smooth; dry; dull white. FLESH thick, firm, white; taste extremely acrid. LATEX white. GILLS subdecurrent, close, forked, white to creamy. STIPE 2.6 cm long, 1-2 cm thick; equal or tapering downward; smooth; dry; white. SPORES white; subglobose; warted and reticulate; 6-8.5 x 6-6.5 μ. Probably POISONOUS.

L. piperatus occurs under hardwoods along the Pacific Coast in the fall. It is suspected of being poisonous.

L. deceptivus Pk. is also white and about the same size, but it often has rusty stains on the white pileus, and its margin is tomentose and at first inrolled. Its edibility is unknown.

Lactarius rufus (Scop. ex Fr.) Fr.   Plate 61

PILEUS 2.5-10 cm in diameter; convex to depressed or funnel-shaped, sometimes umbonate; dry; brownish-red to brick-red. FLESH thin, soft, white or pinkish. LATEX white; taste at first mild, but later extremely acrid. GILLS subdecurrent; close; narrow; occasionally forked; yellowish, becoming reddish. STIPE 5-8 cm long, 6-12 mm thick; smooth; equal; dry; concolorous with pileus. SPORES white; elliptical; reticulate; 7-9 x 5-7 μ. POISONOUS; see below.

L. rufus is widely distributed and grows under conifers. Its latex at first tastes mild, but it becomes so acrid that the burning sensation stays for some time in the mouth. It can cause severe gastrointestinal troubles.

L. rufus should be distinguished from a smaller edible species, **L. subdulcis** (Bull. ex Fr.) S.F. Gray, which is similarly colored, but has a mild-tasting latex and is usually associated with hardwoods.

Lactarius aurantiacus Fr.   Plate 62

PILEUS 3-6 cm in diameter; convex to plane, becoming depressed or umbilicate; smooth; glutinous when wet; reddish-orange. FLESH thin, brittle; orange-buff to pale yellowish. LATEX white; taste mild or slightly bitter. GILLS adnate; close; orange-buff. STIPE 3-6 cm long, 5-10 mm thick; equal or slightly enlarged in middle, tapering at base; concolorous with pileus or paler. SPORES pale yellow, ellipsoid to subglobose, coarsely warted and reticulate, amyloid, 8-10 x 6.5-8 μ. EDIBLE.

L. aurantiacus is common along the West Coast. This very colorful, shining fungus fruits abundantly under conifers in the fall and winter. After heavy rainfall the gluten may be washed away.

Lactarius necator (Pers. ex Fr.) Lund.

PILEUS 5-14 cm in diameter; convex to depressed or umbilicate; viscid; smooth; margin inrolled and usually hairy; olive-brown with darker disc. FLESH firm, whitish. LATEX white; taste acrid. GILLS slightly decurrent; narrow; close; pale yellow, becoming black or brownish when bruised. STIPE 2.5-6 cm long, 1-2.5 cm thick; smooth; equal; viscid; olive-brown, sometimes with darker spots. SPORES buff; subglobose; warted and reticulate; 7-9 x 5.5-7 μ. EDIBILITY unknown.

L. necator is a rather unattractive fungus, with its dull colors, and not likely to appeal to the mushroom hunter. It usually occurs under conifers but sometimes appears in mixed woods.

Lactarius chrysorheus Fr.

PILEUS 4-10 cm in diamter; convex to depressed or umbilicate; slightly viscid when wet; smooth; margin inrolled, often becoming wavy; sometimes azonate, but usually zoned

with yellowish, pinkish, and buff bands. FLESH firm, white.
LATEX white, quickly changing to deep yellow; taste acrid.
GILLS adnate; narrow; close; white, becoming dingy yellow
or pale reddish-brown. STIPE 4-7 cm long, 1-2.5 cm thick;
smooth; equal; stuffed, then hollow; white at first, becoming
yellowish-buff in age. SPORES pale buff; elliptical; reticulate; 6-9 x 5-7 μ. POISONOUS.

L. chrysorheus is a very common, widespread mushroom
that grows under both conifers and hardwoods. The quick
change in the latex from white to yellow is an important
field character.

A similar mushroom, **L. zonarius** (Secr.) Fr., is also
zoned, but its bands are orange and reddish-orange, and its
latex is white and unchanging. It occurs in spring and fall in
the Rocky Mountains and along the Pacific Coast. Its
edibility is unknown.

Lactarius scrobiculatus (Scop. ex Fr.) Fr.

PILEUS 5-15 cm in diameter; convex to depressed; smooth;
viscid; pale yellow to ochraceous or yellowish-orange, sometimes with concentric zones. FLESH firm; white; stains
yellow. LATEX scanty; white, stains bright yellow; taste
acrid. GILLS adnate, slightly decurrent; crowded; narrow;
sometimes forked; white or yellowish, staining dark yellow.
STIPE 3-6 cm long, 1-3 cm thick; hollow; equal; smooth; dry;
concolorous with pileus or paler, with depressed brighter
orange spots. SPORES white; subglobose to ellipsoid; warted
and reticulate; amyloid; 7-9 x 6-7.5 μ. POISONOUS.

L. scrobiculatus derives its name from the *scrobiculations,*
or pits, on the stipe which serve to distinguish it from several
similarly colored species of *Lactarius*. It is common in the
western states and provinces and occurs under conifers.

L. resimus Fr. is similar to *L. scrobiculatus* but has a
much paler pileus at first and lacks the depressions on the
stipe, but its latex also changes from white to sulphur-
yellow. Its edibility is unknown.

Lactarius uvidus (Fr.) Fr.   Plate 63

PILEUS 2.5-9 cm in diameter; convex to depressed, sometimes umbonate; viscid; smooth; brownish-gray with a lilac
tinge. FLESH white, becoming lilac when bruised. LATEX
white, quickly changing to lilac or violet; taste acrid. GILLS
adnate to subdecurrent; close; narrow; whitish, becoming
lilac where bruised. STIPE 2.5-7 cm long, 6-12 mm thick;
smooth; viscid; equal; stuffed, becoming hollow; whitish to
dingy-yellowish. SPORES white; ellipsoid to subglobose;
warted and reticulate; 7-12 x 6-8.5 μ. POISONOUS.

L. uvidus grows on the ground in very wet places, often
among moss, under conifers in the Rocky Mountains and
along the Pacific Coast from northern California to Canada. Its purplish latex and flesh make it easily recognizable.

Lactarius mucidus Burl.

PILEUS 2.5-8 cm in diameter; convex to depressed or
infundibuliform; smooth; viscid; azonate; gray to brownish-
gray. FLESH soft; white or grayish. LATEX white, becoming
pale grayish-green or bluish. GILLS adnate; close; narrow;
white, becoming bluish-gray. STIPE 2-6 cm long, 8-10 mm
thick; smooth; viscid; equal; concolorous with pileus.
SPORES white; elliptical to subglobose; reticulate; amyloid;
7.5-10 x 6-8 μ. INEDIBLE; see below.

L. mucidus is a common species found under conifers in
the western states and provinces. It is reported to be
nonpoisonous, but its acrid taste would make it unpalatable.

A somewhat similar mushroom, **L. trivialis** Fr., also has
a dull-colored pileus with a purplish tinge, but its spores are
yellowish. It is suspected of being poisonous.

Lactarius sanguifluus Fr.

PILEUS 6-12 cm in diameter; convex or depressed; smooth;
viscid; zoned with concentic bands of orange and paler
orange. FLESH firm; buff to orange, staining green in time.

LATEX blood-red. GILLS adnate to subdecurrent; pale purplish-red, becoming greenish-stained in age. STIPE 2-5 cm
long, 1-3 cm thick; hollow; tapered at base; pale orange.
SPORES pale yellow; elliptical; reticulate; amyloid; 7.5-8 x
6.5-8 μ. EDIBLE and choice.

L. sanguifluus is distinguished by its blood-red latex and a
characteristic reddish or purplish sheen on the gills. This
highly desirable edible mushroom is known only in the
western states and provinces where it grows under conifers.

A widely distributed species, **L. deliciosus** (L. ex Fr.) S.F.
Gray (Plate 64), is also associated with conifers and is more
common. It has bright orange gills and latex and buff-
colored spores, 8-11 x 7-9 μ. It is also edible, but we
consider *L. sanguifluus* to be superior. They both develop
green stains in all parts in time, or when the latex is exposed.

FAMILY AMANITACEAE

The family Amanitaceae is characterized by smooth white
spores, free gills, and the presence of a universal veil that
surrounds the young button, leaving a volva, commonly
called a cup, at the base of the stipe when the mushroom
expands. There are only two genera, *Limacella* and *Amanita*.

Limacella is rare and can be distinguished by a slimy,
gelatinous veil; *Amanita* has a membranous veil.

Genus *Amanita*

It is extremely important that anyone collecting mushrooms for the table recognize members of *Amanita* (Fig.
22). It contains the majority of our mostly deadly mushrooms and also some hallucinogenic species. *Amanita* has
cosmopolitan distribution, although many species are delimited to certain areas. They always grow on the ground.
Fortunately, Amanitas are easy to recognize if the collector
is meticulous about digging underground to obtain the
entire specimen. The volva is sometimes powdery, or so

FIG. 22. *Amanita vaginata*

friable that it could be overlooked, but it is an important feature along with white spores and (usually) white gills. There may or may not be an annulus; those species lacking a ring were formerly placed in a separate genus, *Amanitopsis*.

While there are a few edible species, all members of *Amanita* should be avoided unless the specimens are identified by an expert mycologist. The only other gilled genus which is obviously volvate is *Volvariella*, but it has pink spores and no annulus. There have been some cases of poisoning when an *Amanita* button was mistaken for an edible puffball, but the difference is apparent at once if a specimen is cut in half. Instead of only white, homogeneous flesh as found in puffballs, the section of a gilled mushroom will show the embryonic outline of a pileus, gills, and stipe.

The proliferation of *A. phalloides* on the West Coast, particularly in California, in recent years is dismaying. This species was formerly considered to be European and found only rarely in eastern states. It is responsible for most of the deaths or serious illnesses reported. Local mycological societies and toxicology committees make every effort to alert the public to the danger, but still people persist in blindly eating a mushroom that looks "nice."

KEY TO SPECIES OF *AMANITA*

A. Pileus white, olive, or greenish
  1a. Pileus pure white; smooth, floccose,
      or warted **2**
  b. Pileus streaked, greenish-olive; odor of
      raw potato ....*A. phalloides*, p. 198
    2a. Pileus smooth **3**
    b. Pileus floccose or warted **4**
  3a. Pileus white; stipe slender; persistent
      membranous volva........ *A. bisporigera*, p. 199
  b. Pileus white, often with a yellowish
      tinge on disc; stipe thick;
      volva large.................. *A. ocreata*, p. 199
    4a. Pileus with large pyramidal warts; stipe
        tapering downward.. *A. strobiliformis*
                              (not described further)
    b. Pileus floccose, viscid when old;
        stipe short, bulbous, with floccose
        remains of veil ....*A. silvicola*, p. 199
B. Pileus brown, yellowish-brown, reddish-brown,
    or gray
  1a. Pileus grayish-brown with a purplish
      tinge; annulus and
      volva gray .... *A. porphyria*, p. 200
  b. Pileus yellowish-brown; annulus and volva
      with yellowish tints in age..... *A. aspera*, p. 200
  c. Not as above **2**

Amanita phalloides (Fr.) Secr.   Plate 65

PILEUS up to 12 cm in diameter; ovoid when young, expanding to broadly hemispheric with age; margin smooth; viscid, or shining when dry, with a metallic or silky sheen; dull yellow or greenish to olive, streaked with darker fibrils;

surface usually smooth, rarely with fragments of volva. FLESH white or with slightly greenish tinge below cuticle; odor of raw potato. GILLS white; close; free. STIPE up to 20 cm long with bulbous base; relatively slender; smooth; white to greenish-yellow; annulus superior, white, pendant; volva large, free, with margin often lobed. SPORES white; nearly globose; smooth; amyloid; 8-10 μ. VERY POISONOUS.

A. phalloides is found along the Pacific Coast in Oregon and particularly in California and is responsible for most of the mushroom deaths in recent years. This deadly species is usually associated with oaks, and fruits from November through January. Its odor, resembling that of raw potato when fresh, becomes very unpleasant and nauseating when it is dried. The flavor, according to survivors, is excellent. For symptoms of poisoning, see the section in the Introduction on edible and poisonous mushrooms.

An equally toxic mushroom, **A. bisporigera** Atk., occurs in Idaho in the fall and is associated with aspens and birches. It is somewhat similar to *A. phalloides* but is pure white with a slenderer stipe. As its name indicates, it has only 2 spores on each basidium.

A. ocreata Pk. which contains at least two of the same deadly amatoxins, has been reported on the California coast from San Diego County to the San Francisco Bay Area and inland in Madera County. It grows in oak woodlands, and fruits later than the other species, often not starting until February and continuing until May in southern California. At first white, sometimes with a yellowish-tinged disc, it usually becomes buff-color in age. It has a very large, persistent volva. There is a bright yellow color reaction when KOH is applied to the pileus or flesh; this is the only *Amanita* known to show this change.

Amanita silvicola Kauf.

PILEUS up to 10 cm in diameter; convex when young, becoming nearly plane with age; slightly viscid; surface covered with white floccose tissue; white. FLESH white; thick, becoming thin at margin. GILLS white; slightly attached at stipe. STIPE up to 12 cm long; relatively thick; white, with floccose remains of veil which make it difficult to distinguish annulus; volva attached in form of shaggy rings. SPORES white; ellipsoid; smooth; amyloid; 9-12 x 5-6 μ. EDIBILITY unknown.

A. silvicola is found infrequently in the coastal forests of Washington, Oregon, and California in late fall and winter. The pure white color and shaggy pileus and stipe of this beautiful species make it easily identifiable.

Amanita porphyria (Fr.) Secr.

PILEUS up to 7 cm in diameter; convex at first, expanding with age; margin smooth; surface viscid; brownish-gray with a purplish tinge, sometimes with grayish patches. FLESH white; thin; lacks odor. GILLS white; free. STIPE up to 11 cm long with bulbous base; grayish or mottled gray and white; annulus gray, soon collapsing on stipe; volva gray, fragile. SPORES white; globose; smooth; amyloid; 7-9 μ. HALLUCINOGENIC and POISONOUS.

A. porphyria is an uncommon species, but it does occur sparingly in the Rocky Mountains and along the Pacific Coast north to Alaska under conifers or hardwoods in spring and fall.

Amanita aspera (Fr.) Quél.

PILEUS up to 12 cm in diameter; hemispheric or convex, later becoming plane; margin even; surface covered with yellowish scales that become gray in age, on a brown background. FLESH white or yellowish under the cuticle; thick. GILLS free; broad; two tiers of lamellulae; white at first, becoming yellowish near the margin. STIPE up to 15 cm long; tapering upward from a decidedly bulbous base; yellow above annulus, scurfy yellowish zones below; annulus superior, yellow on top, gray below; volva adheres to base of stipe in irregular rings; basal part slowly stains reddish-brown. SPORES white; ellipsoid; smooth; amyloid; 8-10 x 6-7 μ. EDIBILITY questionable.

A. aspera occurs under conifers late in the season along the Pacific Coast. Its edibility is questionable, and because of its brown-staining character it should be carefully differentiated from the edible *A. rubescens* (see below).

Amanita rubescens (Fr.) S.F. Gray   Plate 66

PILEUS up to 12 cm in diameter; hemispheric, expanding with age; margin sometimes slightly striate; whitish at first, becoming reddish-brown with pinkish to red blotches. FLESH white, changing to reddish or reddish-brown when exposed to air. GILLS white; free; becoming reddish-spotted in age. STIPE up to 18 cm long; relatively stout; white above annulus, usually tinged pink below; annulus large, white above with noticeable striations, pinkish below; volva small, fragile, on slightly bulbous base. SPORES white; ellipsoid; smooth; amyloid; 8-9 x 6-7 μ. EDIBLE, but NOT RECOMMENDED.

A. rubescens is not common, but we have found it most frequently along the Pacific Coast, where it grows under oaks. Because of its tendency to bruise red, it is called The Blusher. It is edible and said to be very good, but care should be taken with its identification.

Amanita vaginata (Fr.) Vitt.

PILEUS up to 10 cm in diameter; ovate at first, becoming plane with umbo; margin grooved or striate; viscid; grayish to yellowish-brown, with few or no white patches on surface. FLESH thin; white; lacks odor. GILLS white; free; edges often slightly fimbriate. STIPE up to 18 cm long, usually tapering upward slightly; lacks annulus; surface somewhat fibrillose; whitish to gray or pale tan; volva large, membranous, mostly beneath ground. SPORES white; globose; smooth; amyloid; 8-10 μ. POISONOUS when raw.

A. vaginata is a cosmopolitan species that is frequently found in wooded areas, but it is scattered and usually solitary.

Amanita fulva Schaeff. ex Secr., sometimes listed as **A. vaginata** var. **fulva** Schaeff., is much more orange-brown. It is reportedly edible, but we do not recommend eating any Amanitas.

A. inaurata Secr., which occurs infrequently under conifers, is also considered a variant; in older literature it has been called *Amanitopsis strangulata* (Fr.) Roze. It has a nearly black pileus when young which later becomes a dark or pale gray; the volva is at first white, later becoming gray, and there are also gray warts on the cap.

Amanita gemmata (Fr.) Gill.
Amanita junquillea Quél.

PILEUS up to 7 cm in diameter; convex to plane; margin striate; viscid; bright yellow to creamy-buff with numerous white patches on surface when young. FLESH rather thin; whitish; lacks odor. GILLS whitish; free; edges even. STIPE up to 15 cm long, arising from slightly bulbous base; whitish; smooth above annulus, somewhat scaly below; annulus persistent, superior, white; volva adhering to base, but with a free collar. SPORES white; subglobose to ellipsoid; smooth; nonamyloid; 9-11 x 6-8 μ. EDIBILITY questionable (see below).

A. gemmata is a fairly common and widespread species along the Pacific Coast, where it is found in coniferous and mixed forests. In some literature it is reported to be edible, but should not be eaten because it bears a superficial resemblance to *A. pantherina* (see below), a highly hallucinogenic species that has even caused some deaths. The two are also known to hybridize, but in its true form *A. gemmata* can be distinguished by its proportionately smaller pileus, longer stipe, generally lighter color, and less persistent patches of volval tissue on the cap.

Amanita pantherina (Fr.) Quél.   Plate 67

PILEUS up to 12 cm in diameter; convex when young, becoming plane; margin striate; viscid; creamy or pale yellow to dark brown; numerous pointed white warts on surface. FLESH white; rather thick; odor not distinctive. GILLS white or slightly creamy; close; free. STIPE up to 12 cm long; tapering from moderately thick base; white; smooth above annulus, fibrillose below; annulus median to superior; floccose; white, sometimes with edge yellowish or grayish; volva adhering to bulblike base with a free collar-like margin. SPORES white; smooth; ellipsoid; nonamyloid; 9-13 x 7-9 μ. HALLUCINOGENIC and POISONOUS.

A. pantherina is one of the most common species of *Amanita* in many places along the Pacific Coast and can often be found in profusion under conifers or in mixed woodlands. It occurs in the Rocky Mountains in late summer and autumn, and farther west and south in the Coast Ranges it continues fruiting until spring. In the Pacific Northwest *A. pantherina* hybridizes with *A. gemmata* (see above), and since the latter is reported to be nonpoisonous, there is a variable amount of toxins present. *A. pantherina* contains ibotenic acid and muscimol and possibly other toxins, and has caused fatalities. It produces a highly hallucinogenic state and is intentionally used for recreational purposes in some circles.

Amanita muscaria (Fr.) S.F. Gray   Plate 68

PILEUS up to 20 cm in diameter; hemispheric at first, becoming expanded or plane with age; margin striate; viscid; bright red or orange-yellow (according to variety); numerous white or yellowish pyramidal warts on surface, often arranged in concentric circles. FLESH whitish; fairly thick; lacks odor. GILLS free; white. STIPE up to 20 cm long; tapering upward; white or creamy; fairly smooth above annulus, shaggy below; annulus large, membranous, white; volva arranged more or less in form of concentric scales or rings above bulbous base. SPORES white; ellipsoid; smooth; nonamyloid; 8-11 x 6-8 μ. HALLUCINOGENIC and POISONOUS.

A. muscaria, the Fly Mushroom, derived its name from its use as an agent to stun houseflies (*Musca domestica*). Later a drug was discovered in the mushroom and called *muscarine*. For many years it was believed to be responsible for the effects of intoxication when the fungus was eaten, but now we know the principal toxins are ibotenic acid and muscimol.

This is a very beautiful and quite common mushroom in wooded parts of the West from Alaska south. There are two varieties, *A. muscaria* var. **muscaria**, which is a bright red, and *A. muscaria* var. **formosa**, a yellow-orange form. The latter is much commoner in the eastern states.

Like *A. pantherina*, *A. muscaria* is ingested or smoked for recreational purposes. It seems to be used mostly in California, while *A. pantherina*, or perhaps the hybrid *A. pantherina-gemmata* form, is more popular in the Northwest. Severe hallucinogenic effects requiring hospitalization can result in some cases.

Amanita caesarea (Fr.) Schw.

PILEUS up to 14 cm in diameter; ovoid at first, becoming plane, usually with an umbo at maturity; surface smooth, no warts; viscid; margin striate; reddish-orange in center to yellowish on margin. FLESH yellow; thick. GILLS free or slightly adnexed; broad, close; yellow; edges floccose. STIPE up to 20 cm long, usually tapering upward; smooth or slightly floccose; orange-yellow; annulus large, pendulous, orange-yellow; volva large, white, free from stipe above. SPORES white; ovoid; smooth; 9-9.5 x 6-8 μ. EDIBLE and choice; see below.

A. caesarea is considered choice; it was eaten by the Caesars in ancient Rome, and today it is sold in quantity in European markets. It occurs in wooded areas in the Southwest after summer rains.

The only poisonous mushroom with which *A. caesarea* might be confused is the orange form of *A. muscaria* (see above). However, the smooth pileus with marginal striations, the yellow stipe and gills, and the very large white volva should distinguish *A. caesarea* from the Fly Mushroom.

Amanita calyptroderma Atk. and Ballen   Plate 69

PILEUS up to 20 cm in diameter; convex, becoming plane; slightly viscid; margin becoming tuberculate-striate; yellow-orange, usually with a conspicuous large white patch, occasionally with smaller patches of universal veil on disc. FLESH thick; white to yellowish; no distinctive odor or taste. GILLS white to yellowish; attached to stipe at first, becoming free. STIPE up to 25 cm long; smooth; yellow above annulus, white below; annulus large, pendulant, with upper surface yellow and striated; volva large, white, with free, lobed margin. SPORES white; ellipsoid; smooth; nonamyloid; 8-11 x 5-6 μ. EDIBLE; see below.

A. calyptroderma occurs in great abundance in oak, Madrone, and Douglas Fir forests along the Pacific Coast from southern California to British Columbia about six weeks after the first heavy fall rains; it also occurs in the summer in the Rocky Mountain region. Because it is regarded as a delicacy, it is often collected in quantity, but we caution against eating any *Amanita* until one is thoroughly familiar with the genus.

A. velosa is a somewhat similar mushroom; its large white volval patch on the pileus is comparable in size to that of *A. calyptroderma*. However, its color is a distinctive orange-buff to pinkish-tan, and it has no annulus. Nothing is known of its edibility.

FAMILY HYGROPHORACEAE

A single genus, *Hygrophorus*, is found in the Hygrophor-aceae. It includes many beautiful and brilliantly colored fungi—vivid reds, oranges, yellows, greens, and blues—as well as the more drab gray, brown, or white species. Most have a viscid or glutinous pileus which makes them conspicuous and excellent for color photography.

Genus *Hygrophorus*

Hygrophorus is characterizied by thick adnate or decurrent gills which look and feel waxy when rubbed between the fingers; a stipe confluent with the pileus and usually central, though it may be eccentric in some species; and smooth, white, elliptic, oblong, or globose spores. A veil is often present, but there is no volva. All species of *Hygrophorus* grow on the ground.

The arrangement of the *trama*, or gill tissue, makes it possible to divide this genus into three sections: the trama may be *parallel* (with long, broad hyphal cells); *divergent* (with hyphae extending downward and outward from a median line of parallel cells); or *interwoven* (evenly interlaced throughout). Obviously the examination of these details is a microscopic procedure employed only by professional mycologists or advanced students.

Many of the fleshy species of *Hygrophorus* are edible, but only a few are considered to be very good. One species, *H. conicus*, is suspected of being poisonous.

KEY TO SPECIES OF *HYGROPHORUS*

A. Pileus red, scarlet, or purplish
  1a. Pileus red, streaked with purplish
    fibrils .... *H. capriolarius*, p. 209
   b. Not as above 2
    2a. Pileus dry; depressed or umbilicate;
      scarlet .... *H. miniatus*, p. 211
     b. Pileus moist or viscid 3
  3a. Pileus moist or subviscid,
    red; stipe red, yellowish
    at base .... *H. coccineus*, p. 211
   b. Pileus viscid, deep red; stipe orange
    or yellow .... *H. puniceus*, p. 211
   c. Pileus slightly viscid when moist, reddish
    or scarlet-orange, often with olive tints;
    all parts blacken when
    bruised .... *H. conicus*, p. 212
B. Pileus orange, bright yellow, or dull greenish-yellow
  1a. Pileus dull greenish-yellow, olive-brown
    on disc .... *H. hypothejus*, p. 212
   b. Not as above 2
    2a. Pileus bright yellow or orange-yellow;
      gills yellow .... *H. flavescens*, p. 213
     b. Pileus reddish-orange, bright
      orange, or yellow; gills white
      to yellowish .... *H. speciosus*, p. 213
C. Pileus at first blue or green
  1a. Pileus blue with reddish
    or brown disc; gills bluish,
    becoming yellowish
    .... *H. psittacinus* var. *californicus*, p. 214
   b. Pileus dark green, changing to various shades
    of yellow, pink, flesh-color, etc.; gills
    pale green, becoming reddish or
    yellowish .. *H. psittacinus* var. *psittacinus*, p. 213

D. Pileus pink or pale vinaceous

  1a. Pileus shell pink; conic; stipe up to
16 cm long .... *H. calyptraeformis*, p. 210

   b. Not as above 2

    2a. Pileus rosy-pink, fading to
cream-color in age; spores large,
12-15 x 7-9 μ. .... *H. goetzii*, p. 210

     b. Pileus pink or pale vinaceous
on disc, whitish toward margin; spores
7-10 x 5-6 μ. .... *H. erubescens*, p. 210

E. Pileus shades of gray or brown

  1a. Pileus dry; dark
smoky-brown ....*H. camarophyllus*, p. 214

   b. Pileus viscid 2

    2a. Pileus brown to pinkish-brown;
virgate; margin inrolled and
tomentose .... *H. roseibrunneus*, p. 214

     b. Not as above.........................3

  3a. Pileus yellow-brown or rusty-brown,
becoming almost white
on margin .... *H. bakerensis*, p. 215

   b. Pileus shades of gray or with darker fibrils ...4

    4a. Conspicuously streaked with gray or
black fibrils; stipe with irregular
light and dark bands up to
annulus .... *H. olivaceoalbus*, p. 215

     b. Not as above.........................5

  5a. Margin pale gray, darker on disc;
stipe with minute white or gray,
pointed scales .... *H. tephroleucus*, p. 216

   b. Pileus, gills, and stipe pale
violet-gray .... *H. laetus*, p. 216

   c. Pileus ashy-gray; gills white,
becoming grayish; strong odor
of almonds .... *H. agathosmus*, p. 215

F. Pileus white, pale cream, or pale buff

   1a. Pileus whitish to pale buff, becoming
pinkish-buff; sometimes with ochraceous spots
near margin; odor of dried
peaches .... *H. saxatilis*, p. 216

    b. Not as above 2

      2a. Pileus white to pale cream;
gills ivory or dingy yellow;
glutinous veil .... *H. gliocyclus*, p. 217

       b. Not as above 3

   3a. Pileus and stipe white, dotted with golden-
yellow granules .... *H. chrysodon*, p. 217

    b. Pileus pure white 4

      4a. Pileus broadly convex, becoming
obtuse or plane; viscid; stipe bulbous
when young .... *H. subalpinus*, p. 218

       b. Not as above 5

   5a. Stipe glutinous .... *H. eburneus*, p. 218

    b. Stipe moist to dry .... *H. piceae*, p. 218

Hygrophorus capriolarius (Kalch.) Sacc.

PILEUS 3-6 cm in diameter; convex to expanded, with slight umbo; viscid, soon becoming dry; appressed scales on disc; red, streaked with purplish fibrils. FLESH thick; pallid or reddish. GILLS adnate to decurrent; far apart; ventricose; pallid to pink, becoming concolorous with pileus. STIPE 3-10 cm long, 6-12 mm thick; dry; equal or slightly enlarged in middle; fibrillose; whitish with reddish tinge at first, becoming concolorous with pileus. SPORES white; ellipsoid; smooth; 6.5-8 x 4.5-5 μ. EDIBILITY unknown.

H. capriolarius grows under various species of spruce in the Pacific Coast states and Idaho.

H. purpurascens Fr., which occurs in the same habitat, is very similar in color, but it has a partial veil and a smooth pileus. It is edible.

Hygrophorus erubescens Fr.

PILEUS 5-8 cm in diameter; convex; umbonate at first, becoming plane to depressed or funnel-shaped; fibrillose; viscid; nearly white with a reddish tinge when young, becoming vinaceous. FLESH soft; thick on disc; white, staining yellow. GILLS adnate, becoming decurrent; broad; rather far apart; white when young, becoming shell-pink with reddish stains. STIPE 4-7 cm long; 6-12 mm thick; tapering at base; fibrillose or scaly; brownish-vinaceous with white apex and base. SPORES white; ellipsoid; smooth; 7-10 x 5-6 μ. EDIBLE, but poor.

H. erubescens grows gregariously under conifers in the West. It is edible, but reported to have a poor flavor.

A very good edible, **H. russula** (Fr.) Quél., is similar, but has close or crowded gills, and the entire pileus sometimes becomes wine-red.

Hygrophorus calyptraeformis (Berk. and Br.) Fayod.

PILEUS 2.5-6 cm in diameter; acutely conic at first; slightly viscid when wet; fibrillose; pinkish-red, fading to shell-pink; disc often whitish. FLESH thin to moderately thick; pale pink. GILLS adnate to adnexed; narrow; slightly ventricose; sometimes intervenose; edges wavy and sometimes fringed. STIPE 5.5-16 cm long, 4-8 mm thick; smooth; fragile, often splitting; sometimes twisted; hollow; flesh-color, sometimes lavender-tinged in youth. SPORES white; ellipsoid; smooth; 6.3-8 x 4.5-5 μ. EDIBILITY unknown.

H. calyptraeformis occurs in mixed and coniferous forests in California. It has a distinctive microscopic character, with unusually large sterile cells (cheilocystidia and pleurocystidia) on the edges and sides of the gills.

Another beautiful pastel-pink or pinkish-buff mushroom, **H. pudorinus** Fr., has a fragrant odor; it grows under species of spruce or fir. **H. goetzii** Smith is a snowbank fungus which occurs at higher elevations in California and Oregon; its viscid rosy-pink pileus fades to cream-color in age, the gills and stipe are cream-color, and its spores are quite large, 12-15 x 7-9 μ. The edibility of both species is unknown.

Hygrophorus miniatus Fr.

PILEUS 2-4 cm in diameter; convex to plane, then depressed or umbilicate; smooth; dry or moist; bright scarlet, fading to pale orange or yellow and becoming fibrillose; margin striate. FLESH thin; concolorous with pileus or paler. GILLS adnate, becoming adnexed; broad; rather far apart; paler than pileus, fading to orange or yellow. STIPE 3-5 cm long, 3-4 mm thick; smooth; equal; concolorous with pileus, slowly fading to orange. SPORES white; ellipsoid; smooth; 6-8 x 4-5 μ. EDIBLE, but see below.

H. miniatus is a widespread species which grows gregariously on the ground, often among moss, in deciduous and mixed woods. It is edible, but so small that it takes a great many specimens to make cooking it worthwhile.

Hygrophorus puniceus Fr.

PILEUS 2-7 cm in diameter; bluntly conic at first, becoming umbonate to plane; viscid; deep red, sometimes fading to orange. FLESH thin; reddish-orange to pale orange-yellow. GILLS adnate, becoming adnexed; broad; rather far apart; reddish-orange to yellow. STIPE 2-7 cm long, 1-1.5 cm thick; fibrillose or striate; equal or slightly tapered at base; reddish; soon becoming orange or yellow; base white or yellowish. SPORES white; subellipsoid to oblong; smooth; 8-11 x 4-6 μ. EDIBLE.

H. puniceus is a brilliantly colored mushroom which is conspicuous in hardwood and conifer forests along the Pacific Coast.

A very similar and also edible species, **H. coccineus** Fr. (Plate 70), which occurs in the same habitat, has a moist but not viscid pileus and a shiny red stipe which is usually yellow at the base.

Hygrophorus conicus Fr. Plate 71

PILEUS 2-9 cm in diameter; conic or convex and umbonate; slightly viscid when moist; smooth or streaked with fibrils; usually red or scarlet-orange, but at times yellow, olive, or green; blackens immediately when bruised. FLESH thin; fragile; concolorous with pileus; blackening. GILLS nearly free; broad; close; pallid to olivaceous-yellow; blackening. STIPE 4-12 cm long, 3-10 mm thick; fragile; equal; striate; dry or moist; often twisted; reddish, yellow, or orange; base whitish; blackening. SPORES white; ellipsoid; smooth; 9-12 x 5.5-6.5 μ. May be EDIBLE; see below.

H. conicus occurs in coniferous forests and is very common throughout North America. The persistently conic or umbonate pileus, and the immediate blackening of all parts when handled, make *H. conicus* easily recognizable. It occasionally grows larger than the measurements given above. Older literature lists it as poisonous or suspect, but some well-informed people, such as Dr. Rolf Singer, have eaten it repeatedly with no ill effects.

Hygrophorus hypothejus Fr.

PILEUS 2-8 cm in diameter; convex to expanded and umbonate; sometimes with depressed disc; glutinous; smooth; radially streaked with fibrils; dull greenish-yellow or tawny, disc brown to olive-brown. FLESH thin; yellowish under cuticle, watery-whitish below. GILLS decurrent; rather far apart; at first white, becoming pale yellow to orange-yellow. STIPE 8-16 cm long, 6-12 mm thick; solid; equal or tapering downward; yellowish at apex above indistinct annular zone; glutinous below; olive-brown, yellow, orange, or red. SPORES white; ellipsoid; smooth; 7-9 x 4.5 μ. EDIBLE, but see below.

H. hypothejus is widely distributed, growing gregariously under conifers and in peat bogs. The colors of the pileus are more apparent after the olive-brown gluten disappears. It is edible, although without much substance.

Hygrophorus flavescens (Kauf.) Smith and Hesler

PILEUS 2.5-7 cm in diameter; convex to almost plane; viscid, becoming dry; bright orange-yellow, paler on margin. FLESH thin; yellowish. GILLS adnexed; close; yellow. STIPE 4-7 cm long, 8-12 mm thick; dry or moist; equal; hollow; yellow at apex, orange in middle portion, white at base. SPORES white; ellipsoid; smooth; 7-8 x 4-5 μ. EDIBILITY unknown.

H. flavescens grows in coniferous, deciduous, and mixed forests in Washington, Oregon, and California.

Hygrophorus speciosus Pk.   Plate 72

PILEUS 1-3 cm in diameter; conic to convex, becoming expanded and sometimes umbonate; smooth; glutinous; scarlet to orange-red, shading to yellow near margin. FLESH soft, white or yellowish-tinged beneath cuticle. GILLS decurrent; far apart; narrow; white to yellowish; edges yellowish. STIPE 4-10 cm long, 4-10 mm thick; stuffed; equal or enlarged at base; viscid; whitish or concolorous with pileus. SPORES white; ellipsoid; smooth; 8-10 x 4.5 μ. EDIBILITY unknown.

H. speciosus may grow singly, scattered, or somewhat caespitose under conifers in Oregon, Washington, Idaho, and Arizona.

Hygrophorus psittacinus (Fr.) Fr. var. **psittacinus**

PILEUS 1-3 cm in diameter; conic to convex; becoming campanulate or plane in age; glutinous to viscid; dark green at first, later tawny, pinkish, or orange-pinkish; margin striate in age. FLESH thin; fragile; concolorous with pileus or paler. GILLS adnate; rather far apart; narrow, sometimes ventricose; pale green to buff. STIPE 3-7 cm long, 2-5 mm thick; hollow; equal; viscid; greenish, becoming pinkish, orange, or yellow. SPORES white; ellipsoid; smooth; 6.5-8 x 4-5 μ. EDIBILITY unknown.

H. psittacinus grows under conifers and hardwoods. It is commonly called the Parrot Mushroom because of its range of colors from a parrot-green to various shades of flesh, pink, and yellow in age or when exposed to sunlight.

Another color form, *H. psittacinus* var. **californicus**, known only from northern California, is at first blue with a brown or reddish disc and bluish-buff gills which become yellowish or fawn color.

Hygrophorus camarophyllus (Fr.) Dumée   Plate 73

PILEUS 4-13 cm in diameter; convex to umbonate or plane; moist to viscid when wet; grayish-brown or with darker streaks over all. FLESH firm; thick; white. GILLS adnate to slightly decurrent; close to far apart; whitish or ashy-tinged. STIPE 4-13 cm long, 1-2 cm thick; dry; solid; equal or tapering downward; pruinose at apex; fragile; pale grayish-brown. SPORES white; ellipsoid; smooth; 7-9 x 4-5 μ. EDIBILITY unknown.

H. camarophyllus is a large, handsome mushroom which may occur in abundance under conifers near melting snow-banks in the Pacific Northwest.

H. calophyllus Karst., a related mushroom somewhat similar in color, has a glutinous or viscid pileus and pale pinkish gills. **H. marzuolus** (Fr.) Bres., found near snow-banks in Idaho in late spring, also has a virgate but lighter-colored pileus and more widely spaced gills. The edibility of all three species is unknown.

Hygrophorus roseibrunneus Murr.

PILEUS 2-9 cm in diameter; convex, becoming umbonate or plane; sometimes with raised margin; viscid; brown to pinkish-brown, virgate toward disc. FLESH soft; thick on disc; white. GILLS adnate, soemtimes becoming adnexed or decurrent; crowded; white; partial veil evanescent. STIPE 3-9 cm long, 4-18 mm thick; dry; stuffed or solid; pruinose;

white, becoming yellow at base. SPORES white; ellipsoid; smooth; 7-8 x 3.5-5 μ. EDIBILITY unknown.

H. roseibrunneus grows gregariously under conifers and oaks in California.

Hygrophorus bakerensis Smith and Hesler Plate 74

PILEUS 4-15 cm in diameter; convex to plane, sometimes with elevated margin; glutinous when wet; yellow-brown to rusty-brown, shading to nearly white at margin. FLESH firm; thick; white; strong odor of almonds. GILLS decurrent; close to rather far apart; creamy-white. STIPE 7-14 cm long, 8-25 mm thick; solid; dry; equal or tapering downward; white or pinkish-tan. SPORES white; ellipsoid; smooth; 7-9 x 4.5-5 μ. EDIBILITY unknown.

H. bakerensis is a conspicuous mushroom which is common in the fall in coniferous forests along the Pacific Coast and northern Rocky Mountains. Its brown pileus with a much paler margin distinguishes it from **H. agathosmus** Fr., which also has a strong almond odor but is ashy-gray.

Hygrophorus olivaceoalbus Fr.   Plate 75

PILEUS 3-8 cm in diameter; convex to umbonate or plane; glutinous or viscid; streaked with dark gray to blackish fibrils; margin pale gray. FLESH soft; thick on disc; white. GILLS adnate to decurrent; broad; close to rather far apart; white. STIPE 8-12 cm long, 1-3 cm thick; solid; equal or clavate; with double sheath; outer layer glutinous; inner layer of blackish fibrils, breaking up in age to leave light and dark bands up to membranous annulus; white and pruinose above annulus. SPORES white; ellipsoid; smooth; 9-12 x 5-6 μ. EDIBLE.

H. olivaceoalbus is found scattered, gregarious, or caespitose under Sitka Spruce and Coast Redwood along the Pacific Coast. It also occurs in the Rocky Mountains from Colorado northward. The double sheath and conspicuous bands are distinctive.

Hygrophorus tephroleucus Fr.

PILEUS 1-3 cm in diameter; convex to plane or depressed; fibrillose-scaly; viscid; dark gray, fading to ashy-gray in age. FLESH thin; soft; whitish. GILLS adnate, becoming decurrent; broad; rather far apart; white to creamy. STIPE 4-6 cm long, 2-4 mm thick; equal; solid; white; apex covered with white, pointed scales which become gray; base glutinous at first, becoming dry. SPORES white; ellipsoid; smooth; 8-10 x 4-5 μ. EDIBILITY unknown.

H. tephroleucus grows gregariously under conifers in California, Oregon, and Idaho. The gray scales on the pileus and upper part of the stipe are outstanding characters.

Hygrophorus laetus Fr.

PILEUS 1-3.5 cm in diameter; convex to plane or slightly depressed; viscid; smooth; violet-gray; margin sometimes elevated. FLESH thin; tough; concolorous with pileus or paler; may have slightly fishy odor. GILLS adnate to decurrent; narrow to moderately broad; rather far apart; concolorous with pileus. STIPE 3-12 cm long, 2-4 mm thick; viscid; smooth; pliant; concolorous with pileus or paler. SPORES white; ellipsoid; smooth; 5-7 x 3-4 μ. EDIBILITY unknown.

H. laetus occurs along the Pacific Coast and in Idaho. It has many color variants ranging from olivaceous-orange to tawny-olive or pale pink, and has at times been confused with *H. psittacinus* (see above). Unless the latter is observed when young (with its green or blue coloring), it cannot be differentiated macroscopically from *H. laetus*, because they both become flesh-color or orange-pink upon drying.

Hygrophorus saxatilis Smith and Hesler

PILEUS 3-8 cm in diameter; convex to obtuse, becoming plane, sometimes with low umbo; slightly viscid when young, becoming moist or dry; whitish to pale buff, becoming pinkish-cinnamon, sometimes with ochraceous spots near margin. FLESH soft; thick; pinkish-buff; often has odor of dried peaches. GILLS slightly decurrent; narrow to moderately broad; fragile; rather far apart; yellowish-salmon to light pinkish-cinnamon. STIPE 6-8 cm long, 1-1.5 cm thick; solid; equal or slightly narrowed at base; fibrillose; concolorous with gills. SPORES white; subellipsoid; smooth; 7-9.5 x 4-5 μ. EDIBILITY unknown.

H. saxatilis is so named for its propensity for growing on very dry, rocky ground, but it is also associated with conifers. It fruits in the fall in Oregon, Washington, and Idaho. Its beautiful gills and odor of dried peaches should make this mushroom easily identifiable.

Hygrophorus gliocyclus Fr.

PILEUS 4-9 cm in diameter; convex to expanded, subumbonate; glutinous; smooth; white to pale cream. FLESH firm; thick on disc; white. GILLS adnate to decurrent; rather far apart; whitish to ivory or dingy-yellow. STIPE 3-6 cm long, 8-12 mm thick; equal or slightly ventricose, tapering abruptly at base; lower part with glutinous veil up to annulus; fibrillose; whitish above, dingy cream-color below. SPORES white; ellipsoid; smooth; 8-10 x 4.5-6 μ. EDIBLE, but see below.

H. gliocyclus grows under conifers. It is reported to be edible, but the extremely slimy pileus and stipe make it difficult to collect and prepare for cooking. It occurs under spruce and pine in Idaho, Wyoming, Oregon, and California.

Hygrophorus chrysodon (Fr.) Fr.

PILEUS 3-8 cm in diameter; convex to expanded, subumbonate; viscid when wet; white with golden-yellow granules on margin or over surface; margin floccose and inrolled. FLESH soft; thick; white. GILLS decurrent; broad; far apart; intervenose; white; edges sometimes yellow. STIPE 3-8 cm long, 6-12 mm thick; stuffed; viscid; equal; white; apex with yellow flocculose zone. SPORES white; ellipsoid; smooth; 7-9 x 3.5-4.5 μ. EDIBLE and choice.

H. chrysodon grows singly to gregariously under conifers over the Northern Hemisphere. In the West it is known to occur on the Pacific Coast and throughout the Rocky Mountains south to New Mexico. It is easily identified by the golden granules on the pileus and apex of the stipe.

Hygrophorus subalpinus A.H. Smith

PILEUS 4-6 cm in diameter; convex, becoming plane, sometimes umbonate; viscid; white; margin often with veil fragments. FLESH soft; thick; white. GILLS decurrent; narrow; close; white. STIPE 3-4 cm long, 1-2 cm thick at apex; base bulbous when young, later becoming almost equal; inferior membranous annulus; white. SPORES white; ellipsoid; smooth; 8-10 x 4.5-5 μ. EDIBLE; but see below.

H. subalpinus is known in the West from the Pacific Coast states, Idaho, Colorado, and Wyoming. It is gregarious under conifers, often near snowbanks. It is edible, but reported to have little flavor.

Hygrophorus piceae Kühn. and Romag.

PILEUS 1-4 cm in diameter; convex, becoming plane; sometimes with elevated margin; smooth; viscid; white. FLESH soft; white. GILLS adnate, becoming decurrent; broad; white, becoming pinkish-buff. STIPE 3-5 cm long, 3-5 mm thick; moist or dry; tapering downward; upper part fibrillose, lower half smooth; hollow; white. SPORES white; ellipsoid; smooth; 6-8 x 4-5 μ. EDIBILITY unknown.

H. piceae often grows under species of spruce in the West. It can be distinguished by its moist or dry stipe.

Another pure white species, **H. eburneus** Fr., is somewhat larger. Its pileus is usually 2-7 cm, but may measure up to 10 cm in diameter, and the stipe is glutinous. **H. russocoriaceus** Berk. and Miller is another small, white species that is common in fall and winter along the Pacific Coast. It is recognizable by its strong odor of cedar. Nothing is known about the edibility of these species.

FAMILY LEPIOTACEAE

The Lepiotaceae was formerly considered to consist of a single genus, *Lepiota*, but some mycologists have felt that a number of changes were needed. First, there was the matter of spore color, because one species had green spores, while the rest have white, cream, or pale buff spores. It was generally conceded that the green-spored member belonged in a separate genus, and it is now called *Chlorophyllum molybdites*.

Because of certain microscopic differences, some of the remaining members of *Lepiota* have also been transferred to new genera (*Leucocoprinus*, *Leucoagaricus*, and *Macrolepiota*) by Dr. Rolf Singer. However, since it is a highly technical distinction, for convenience we are following Shaffer's (1968) classification and treating the genus *Lepiota* in the broad sense.

Genus *Chlorophyllum*

The unique green spores are the most distinctive feature in members of *Chlorophyllum*. However, the spores do not assume that color until maturity; thus *C. molybdites* has sometimes been mistaken for *Lepiota rachodes*, an edible species (see below). A fresh spore print reveals bright grass-green spores, but this color tends to fade in time to a slate-green.

Chlorophyllum molybdites (Meyer ex Fr.) Mass.
Lepiota morgani (Pk.) Sacc. Plate 76

PILEUS 10-25 cm in diameter; hemispheric when young, later becoming plane; surface dry and smooth at first, soon breaking up into scaly patches, which are later lost. FLESH thick; white; sometimes becomes reddish if bruised; odor and taste not distinctive. GILLS whitish at first, soon turning greenish, then olive. STIPE up to 20 cm in length, 3 cm thick; somewhat enlarged at base; smooth; whitish, staining brown with age; annulus large, thick, persistent. SPORES green; ellipsoid; smooth; 9-12.5 x 6-8 μ. POISONOUS.

C. molybdites is a subtropical to tropical species known from most of the warmer parts of the world. While not uncommon in parts of the East, it is known principally in the West from southern California and Arizona, where it tends to come up on lawns during the summer. Its northern-most recorded occurrence is Oroville, California, where it was the cause of a poisoning.

Genus *Lepiota*

Members of *Lepiota* are characterized by free gills, a central stipe easily separable from the pileus, an annulus, and no volva.

KEY TO SPECIES OF *LEPIOTA*

1a. Pileus scaly 2
  b. Pileus smooth; white or grayish . . *L. naucina*, p. 221
    2a. Gills or flesh change color when bruised 3
      b. Color unchanging 4
3a. Pileus large, robust; concentric dark brown
    scales . . . . *L. rachodes*, p. 221
  b. Pileus small; reddish-brown scales; parts stain
    bright flame red instantly . . *L. flammeatincta*, p. 224
    4a. Pileus minutely scaly; gray or black
      on disc . . . . *L. atrodisca*, p. 222
      b. Not as above 5
5a. Pileus with yellow or brownish scales
    on white background; stipe with
    floccose sheath . . . . *L. clypeolaria*, p. 223
  b. Pileus and scales some shade of brown or red 6
    6a. Pileus small; reddish-brown scales;
      strong odor . . . . *L. cristata*, p. 223
      b. Pileus moderately large;
      reddish flattened scales; darker
      on disc . . . . *L. rubrotincta*, p. 224

Lepiota naucina (Fr.) Quél.   Plate 77
Leucoagaricus naucinus (Fr.) Sing.

PILEUS 5-10 cm in diameter; ovoid when young to broadly
convex with age; surface dry; smooth; white at first, becoming somewhat grayish later; disc sometimes becomes light
brown. FLESH thick; white; no distinct odor or taste. GILLS
white when young, becoming pinkish with age. STIPE 5-10
cm long, 1.5 cm in diameter, sometimes slightly enlarged
basally; white; annulus thick, persistent. SPORES white;
ellipsoid; smooth; 7-9 x 5-6 μ. EDIBLE, except in a gray
variant (see below).

   L. naucina occurs throughout the United States and
southern Canada principally in open areas such as pastures,
lawns, and roadsides, although it is occasionally found in
open woodlands and even coniferous forests. It is considered
quite edible—but because of its similarity to certain deadly
members of the genus *Amanita*, be sure of identification!
There is a gray form that is reported to be poisonous.

Lepiota rachodes (Vitt.) Quél.
Macrolepiota rachodes (Vitt.) Sing.

PILEUS 8-20 cm in diameter; almost round at first, then
becoming hemispheric and finally nearly plane; surface dry,
with cuticle broken in series of large brown scales often
arranged into concentric rings with some fibrillose, pinkish-
white flesh exposed between. FLESH thick; white, stains
reddish; lacks distinctive odor or taste. GILLS white to pale
buff. STIPE up to 20 cm long, 2-3 cm thick; enlarged at base;
smooth; white, stains with age. SPORES white; ellipsoid;
smooth; 10-12 x 5-7 μ. EDIBLE and choice, but see below.

   L. rachodes (Fig. 23) is a large and often abundant species
in parts of the West, especially California. It is usually
gregarious to caespitose and is often found in the open
around corrals or where straw or manure occurs. We have
found it, however, growing under oaks and in coniferous
areas.

FIG. 23. *Lepiota rachodes.*

L. rachodes is among the choicest of edible mushrooms. In the Southwest there is a danger of confusing it with the poisonous, green-spored **L. molybdites**, but this species has not been reported in the Rocky Mountains or Pacific Northwest; hence the likelihood of confusion in these areas seems minimal.

Lepiota atrodisca Zeller

PILEUS up to 5 cm in diameter; hemispheric at first, then almost plane; thin; surface dry, minutely scaly; dark gray to black in center of disc, often fading to white near margin. FLESH thin; white. GILLS white, rather far apart. STIPE up to 6.5 cm long; slender; white; annulus with black margin; persistent. SPORES white; ellipsoid; smooth; 7-8 x 4-5.5 μ. EDIBILITY unknown.

L. atrodisca is found in early winter in conifer forests along the Pacific Coast. This small, delicate species with its distinctively colored pileus and black-rimmed annulus often grows with *L. cristata* and *L. flammeatincta* (see below) in dark, moist areas under species of spruce and fir.

L. sequoiarum is another small species that occurs in the coastal forests of northern California and Oregon. Unlike *L. atrodisca*, the pileus is white except for a tinge of yellow in the center and the annulus, while persistent, tends to collapse on the stipe. Its edibility is unknown.

Lepiota clypeolaria (Bull. ex Fr.) Kumm.   Plate 78

PILEUS 3-8 cm in diameter; ovoid when young, expanding to plane, sometimes with slight umbo; surface dry and scaly; disc dark brown; scales yellow or brownish, with light flesh exposed between them; area between scales often yellowish, especially near margin, which is ragged, often rimmed with loose, floccose patches. FLESH white to yellowish; thin; lacks distinct odor or taste. GILLS white. STIPE up to 10 cm long, 2.5-8 mm in diameter; tapering upward; surface with shaggy sheath; white to yellowish; annulus soon becoming floccose and losing identity. SPORES white; fusiform; smooth; 10-18 x 4-6 μ. POISONOUS.

L. clypeolaria is quite common in autumn and early winter in coniferous forests along the Pacific Coast and in early autumn in the Rocky Mountains; it is usually found in small groups. The shaggy stipe and frequent presence of numerous floccose, yellow patches on the margin of the pileus and on the stipe are very distinctive. It contains gastrointestinal irritants.

Lepiota cristata (Fr.) Kumm.

PILEUS up to 5 cm in diameter; ovate at first, expanding to plane with umbo usually present; white with reddish-brown scales arranged in concentric rings; umbo reddish-brown. FLESH white; fragile; odor fishy, unpleasant. GILLS white; close; margin irregular. STIPE up to 10 cm long; slender; smooth; white to slightly pinkish-brown; annulus soft, white, and tending to disappear. SPORES white; very irregular in shape, varying from elliptical to almost triangular; 5.7-8.3 x 3.5-4.7 μ. EDIBILITY unknown.

L. cristata is common early in the autumn in Pacific coastal forests, where it grows in deep duff.

Lepiota flammeatincta Kauf.   Plate 79

PILEUS up to 6.5 cm in diameter; ovoid when young, becoming broadly convex to nearly plane with age; surface dry with numerous reddish-brown scales between which white flesh is seen, except on disc; turns red within seconds when even slightly bruised. FLESH white; thin. GILLS white, do not change color. STIPE up to 10 cm long; slender; equal; white; turns bright red when handled; annulus persistent, turns red when bruised. SPORES white; ellipsoid; smooth; 9-11 x 4.5-6 μ. EDIBILITY unknown.

L. flammeatincta occurs in autumn and early winter in coastal coniferous forests of northern California and Oregon in damp, shaded situations beneath Douglas Fir, Sitka Spruce, Western Hemlock, and Lowland Fir. Though small, this is one of the most spectacular of gilled fungi because of the almost instantaneous change to bright red on the part of the stipe below the annulus and on the surface of the pileus when it is touched. This red coloration lasts only a short while, then rapidly fades. No color change occurs on either the gills or the stipe above the annulus.

In **L. roseifolia**, which is less common on the Pacific Coast, the gills but not the pileus or stipe rapidly turn pink when bruised, then fade. Its edibility is unknown.

Lepiota rubrotincta Pk.   Plate 80
Leucoagaricus rubrotinctus (Pk.) Sing.

PILEUS 4.5-6.5 cm in diameter; plane, slightly umbonate; surface dry, radially fibrillose, splitting toward margin; minute scales present, more prominent on disc; reddish-brown on disc, becoming pale pink to light buff toward margin. FLESH thick; fragile; white; odor not distinctive. GILLS white, unchanging; free; thin; close; fragile. STIPE 8-10 cm long; slender; hollow at maturity; white; annulus present, superior; white. SPORES white; ellipsoid; smooth; 8.3-9 x 5-6.5 μ. EDIBILITY unknown.

L. rubrotincta is common on the Pacific Coast, especially in deep Coast Redwood duff. It appears relatively early in the season, often with *L. cristata* and *L. clypeolaria.*

FAMILY TRICHOLOMATACEAE

This is an articial family consisting of those gilled fungi which cannot be included in the white-spored families Lepiotaceae, Russulaceae, Amanitaceae, and Hygrophoraceae. Some mycologists refer to it as the "dumping ground" because of the diversity of its members and their characteristics. The spores of most genera are white, but some are lilac, pink, or flesh-colored. They may be stipitate or sessile and may occur on wood or on the ground. The gills are attached and not waxy. The only other white-spored family with attached gills is the Hygrophoraceae, but the waxy feeling and appearance of their gills is distinctive.

While there are several edible species in this family and relatively few poisonous ones, there are not many that are considered outstanding. Those most highly regarded are *Pleurotus ostreatus, Marasmius oreades, Tricholoma flavo-virens,* and *Armillariella mellea.*

KEY TO GENERA OF TRICHOLOMATACEAE

226   *Basidiomycetes: Gilled (Agarics)*

Genus *Asterophora*

Asterophora is a parasitic genus whose members are usually found on *Lactarius* or *Russula*. This fungus may at first resemble a group of small puffballs on the pileus of the host, but as the stipe elongates, it proves to be an agaric, although in some cases the gills fail to develop. The host often looks unappealing because it is usually badly decomposed.

Asterophora lycoperdoides (Bull. ex Mérat)
Ditmar ex Fr.   Plate 81

PILEUS 1-3 cm in diameter, nearly globose at first; dry; white. FLESH firm; pallid; odor and taste of meal. GILLS adnate; distant; pallid. STIPE 2-5 cm long; equal or clavate; white. SPORES brown; globose; spiny; 12-18 μ. EDIBILITY unknown.

Genus *Cystoderma*

Cystoderma was formerly included in *Lepiota*, but it was reclassified as a separate genus because the gills are not entirely free. It is characterized by a granular covering of the pileus and stipe below the annulus. Sometimes the annulus may be rather indistinct, and a volva is lacking. Spores are white and may or may not be amyloid. Members of *Cystoderma* occur on the ground, usually among conifer needles, or on wood. It is a small genus with only 14 species known from North America.

Cystoderma fallax A.H. Smith and Sing.   Plate 82

PILEUS 2-5 cm in diameter; convex to campanulate; dry and covered with granular scales; margin at first appendiculate with veil remnants; rusty-brown. FLESH thick at disc; white or somewhat reddish; odor sometimes unpleasant. GILLS adnate; close; pinkish-buff. STIPE 3-6 cm long, 3-7 mm thick; equal or clavate at base; sheathed with heavy granular covering, concolorous with pileus up to wide membranous annulus, smooth and paler above. SPORES white; ellipsoid; smooth; 3.6-4.5 x 2.8-3.5 μ. EDIBILITY unknown.

C. fallax occurs under conifers, often among moss, in the coastal states, Idaho, and Colorado. It is a striking little mushroom with its bright coloration.

A larger species, **C. cinnabarinum** (Alb. and Schw. ex Secr.) Fayod, has a bright red to orange-brown pileus and white to creamy gills with fimbriate edges. Its annulus is evanescent, and the spores are nonamyloid. A much less noticeable species, **C. amianthinum** (Scop. ex Fr.) Fayod, has a yellowish-brown to yellowish-orange pileus, 1.5-4.5 cm in diameter, and a relatively long stipe. Its annulus is also evanescent, and the spores are amyloid. Their edibility is unknown.

Genera *Armillaria* and *Armillariella*

Armillaria contains many unrelated species and, although several have been transferred to other genera such as *Pleurotus*, *Clitocybe*, and *Tricholoma*, many are still in doubt. *Armillaria* is presently defined as a genus with white, nonamyloid spores; attached gills; a filamentous cuticle; and an annulus but no volva.

Its relative, *Armillariella*, differs in having amyloid spores and a pileus cuticle of appressed hyphae. It is parasitic on wood or underground roots, while most members of *Armillaria* grow on the ground. None is known to be poisonous, but edibility has not been established in all.

Armillaria albolanaripes Atk.

PILEUS 5-12 cm in diameter; subconic to convex when young, often becoming plane; margin sometimes turning up; surface slightly viscid, minutely scaly; yellow, becoming brownish in center. FLESH thick; white or slightly yellowish; lacks distinctive odor or taste. GILLS sinuate to adnexed; changing from white to deep yellow with age. STIPE up to 7.5 cm long; moderately thick; lower half with numerous scales arranged more or less in concentric zones terminating at the annulus, smooth above annulus; white at first, soon becoming yellowish. SPORES white; ellipsoid; smooth; non-amyloid; 5-7 x 3-4.5 μ. EDIBILITY unknown.

A. albolanaripes is a fairly common species both in the Rocky Mountains and along the Pacific Coast from California northward, occurring in the spring as well as the fall. It is commonly associated with conifers and usually grows singly in the humus.

Armillaria ponderosa (Pk.) Sacc.

PILEUS 7-22 cm in diameter; broadly convex to nearly plane; margin inrolled when young; surface smooth, slightly viscid when moist; white except toward center, where orange-brown. FLESH white; firm; thick; odor of pine. GILLS white at first, staining rusty or pinkish-brown; close; hardly reaching stipe. STIPE 10-15 cm long; thick, often tapering toward base; white above annulus, often staining rusty or reddish-brown below; annulus distinct. SPORES white; subglobose; smooth; nonamyloid; 5.5-7.2 x 4.7-5.6 μ. EDIBILITY unknown.

A. ponderosa is most often found in hemlock forests or where Beach Pine grows, on the Pacific Coast; in the Rocky Mountains it is found in Lodgepole Pine, larch, and Western White Pine forests. Its white color, its thick, tapering stipe with an annulus, and its strong pine odor are distinguishing marks.

Armillaria zelleri Stuntz and A.H. Smith

PILEUS up to 10 cm in diameter; margin wrinkled, becoming plane with age; surface somewhat scaly, viscid; orange-brown. FLESH white, becomes brown when bruised; thick; odor and taste like meal. GILLS white when young, soon spotting brown; adnate to adnexed. STIPE 5-10 cm long; up to 2.5 cm in diameter close to pileus, tapering downward; whitish above annulus, marked by orange scales below; annulus white above, orange below. SPORES white; ellipsoid; smooth; nonamyloid; 4-5.5 x 3-4 μ. EDIBILITY unknown.

A. zelleri is associated with pines and most frequently grows on sandy soil. The species was named in 1948 by Stuntz and Smith from specimens collected in Washington and Oregon. Subsequently it has been found in California and the Rocky Mountains. Its bright orange color, annulus, and stipe tapering to a point basally are good field characters.

Armillariella mellea (Vahl ex Fr.) Karst.

PILEUS 4-9 cm in diameter; convex to nearly plane; margin sometimes striate; surface usually with few dark scales; honey-yellow, often rusty-tinged. FLESH moderately thin; whitish; taste sometimes slightly acrid. GILLS adnate to decurrent; creamy white, developing rusty-brown stains. STIPE up to 20 cm long; equal or somewhat enlarged basally; finely fibrillose; becoming hollow with age; pallid above annulus, yellowish to rusty below; annulus white. SPORES white; ellipsoid; smooth; amyloid; 7-9.5 x 5-6.5 M. EDIBLE, but see below.

A. mellea (Fig. 24), commonly called the Honey Mushroom, is a very widespread species or complex of species. The fruiting bodies often grow in clumps of considerable size at the base of living trees and stumps. As a parasite it causes great economic loss in fruit orchards and to ornamental shrubs. It is attached to the living roots by thick, compacted strands of mycelium called *rhizomorphs*. *A.*

FIG. 24. *Armillariella mellea.*

mellea is highly regarded as an edible by many, but only the young buttons should be used. It is not likely to be confused with any other fungus, yet it often eludes quick identification until one becomes familiar with it. The characteristic growth in dense clumps, the honey-colored pileus with a few dark spots, the white spores, and the presence of an annulus, at least when young, are features to look for. There are a great many color forms, varying from almost white to nearly black. Dr. Rolf Singer estimates that this complex group, when thoroughly studied, will eventually be broken up into over 20 species or varieties.

Genus *Mycena*

Mycena is a sometimes confusing genus, containing a large number of species of small, fleshy, white-spored gilled fungi which have a conic or convex pileus with a straight margin. The gills are adnate to decurrent on the slender, tubular, cartilaginous stipe. Some species are terrestrial, but many grow on dead wood or other types of decaying plants. They are abundant and are among the first fungi to appear after the fall rains as well as in the spring in our western mountains after the snow melts, but their small size and fragility make them of little interest to the mushroom hunter. However, many are beautifully colored and thus a delight to see. Dr. A. H. Smith's 1947 monograph on the genus lists well over 230 species from North America.

KEY TO SPECIES OF *MYCENA*

1a. Growing near snowbanks in western
mountains; lower stipe densely covered with
white, woolly hairs .... *M. overholtzii*, p. 238
  b. Not associated with snow 2
   2a. Flesh exudes a white or colored
  juice when cut 3
    b. Not as above 4

3a. Juice blood-red .... *M. haematopus*, p. 235
  b. Juice white .... *M. galopus*, p. 236
    4a. Flesh has noticeable odor when bruised..... 5
      b. Not as above 6
5a. Odor alkaline; pileus grayish-brown with
    paler margin................. *M. alcalina*, p. 233
  b. Odor of radish; entire fruiting body
    some shade of purple............. *M. pura*, p. 237
    6a. Gill edges colored 7
      b. Not as above 8
7a. Gill edges bright
    orange .... *M. aurantiomarginata*, p. 234
  b. Gill edges dark purple .... *M. purpureofusca*, p. 238
    8a. Pileus gray or grayish-brown 9
      b. Not as above 10
9a. Pileus up to 2 cm in diameter; conspicuously
    striate to disc.............. *M. occidentalis*, p. 237
  b. Pileus up to 7 cm in diameter; darker
    on disc .... *M. galericulata*, p. 237
    10a. Pileus yellow; gills pale
      lilac.................... *M. lilacifolia*, p. 236
      b. Not as above 11
11a. Pileus and stipe viscid; yellowish
    or grayish-green; paler
    on margin................. *M. epipterygia*, p. 235
  b. Not as above 12
    12a. Pileus pinkish-red, fading to
      whitish .... *M. amabilissima*, p. 234
      b. Pileus pale pink .... *M. rosella*, p. 238

Mycena alcalina (Fr.) Quél.

PILEUS up to 3 cm in diameter; conic or convex to nearly campanulate, sometimes with umbo; slightly striate when young, becoming noticeably so with age; surface with powdery bloom when young; dark grayish-brown, soon becoming pallid. FLESH white to grayish; odor strongly alkaline. GILLS adnate; fairly far apart; white to grayish. STIPE up to 8 cm long; slender; smooth, covered with bloom at first; white near pileus, otherwise grayish-brown. SPORES white; ellipsoid; smooth; amyloid; 7.5-10 x 4.5-7 μ. EDIBILITY unknown.

M. alcalina is one of the most common Mycenas in North America occurring in coniferous forests on rotten wood and humus. It changes markedly in color from dark to light with age, but the strong alkaline odor of the flesh when crushed is a ready means of identification.

Mycena amabilissima A.H. Smith

PILEUS up to 2 cm in diameter; conic to campanulate; surface smooth, sometimes translucent or striate; pinkish-red fading to whitish. FLESH pinkish to white; lacks distinct odor. GILLS adnate; rather far apart; connected by veins; white to pinkish-red. STIPE up to 5 cm long; slender; smooth but dull; often pinkish when young, becoming yellowish to white. SPORES white; ellipsoid; smooth; nonamyloid; 7-9 x 3-4 μ. EDIBILITY unknown.

M. amabilissima can be found in the fall under conifers and is widely distributed throughout our northern forests.

Mycena aurantiomarginata (Fr.) Quél.

PILEUS up to 2 cm in diameter; conic to campanulate, becoming almost plane; striate; dark olive in center, becoming orange toward margin. FLESH pale yellowish; odor not distinctive. GILLS adnate; pale yellow to olive, edges bright orange. STIPE up to 8 cm long; slender; grayish-olive to orange, becoming orange near stipe; base with bright orange-yellow hairs. SPORES white; ellipsoid; smooth; amyloid; 7-9 x 4-5 μ. EDIBILITY unknown.

M. aurantiomarginata is found under conifers along the Pacific Coast. Fruiting bodies are widely scattered, although an abundant growth can occasionally be found. The bright orange edges of the gills are bound to attract anyone who finds this colorful little fungus.

Mycena epipterygia (Fr.) S.F. Gray

PILEUS up to 2 cm in diameter; ovoid at first, then conic; striate at maturity; viscid; grayish-green to yellow, becoming white on margin. FLESH yellowish; sometimes with faintly fragrant odor. GILLS adnate; fairly far apart; pallid to yellowish. STIPE up to 8 cm long; slender; viscid; yellow. SPORES white; ovoid; smooth; amyloid; 8-10 x 5-6 μ. EDIBILITY unknown.

M. epipterygia occurs on humus in coniferous forests over much of the United States and southern Canada. The grayish or yellowish-green pileus combined with the yellow stipe are good identifying characters.

The closely related **M. epipterygioides** is somewhat similar in appearance but has gills that become red-spotted and the stipe is often reddish at the base.

Mycena haematopus (Fr.) Quél.

PILEUS up to 4 cm in diameter; ovoid to conic or campanulate; margin crenate or scalloped; surface powdery at first, soon polished; moist, appearing striate; reddish-brown, often with vinaceous tinge, paler toward margin. FLESH pallid to vinaceous; odor not distinctive, taste often bitter; exuding red latex when cut. GILLS adnate; whitish. STIPE up to 9 cm long; slender; tubular; dark reddish-brown; powdery when young; hairlike mass at base; exuding red juice when cut. SPORES white; ellipsoid; smooth; amyloid; 8-11 x 5-7 μ. EDIBILITY unknown.

M. haematopus is a common species in forested areas of North America from Alaska southward. It grows on stumps and rotten logs, often in caespitose groups, and is readily recognizable by its reddish color and the dark red latex it exudes when cut.

M. haematopus var. **marginata** is very similar, but the edges of the gills are red instead of pallid.

Mycena galopus (Fr.) Quél.

PILEUS up to 3 cm in diameter; ovoid to conic when young, becoming campanulate with margin slightly upturned; translucent and striate; surface minutely powdery; blackish in center, becoming white toward margin. FLESH grayish-white; lacks distinct odor. GILLS adnate; white or slightly grayish. STIPE up to 10 cm long; slender; pale gray at apex, becoming almost black at base; exuding white latex when broken. SPORES white; ellipsoid; smooth; sometimes slightly amyloid; 9-13 x 5-6.5 μ. EDIBILITY unknown.

M. galopus is one of the most abundant species of *Mycena* on the Pacific Coast. The grayish to almost black pileus, combined with a stipe that exudes a white, milklike latex, readily identifies it.

Mycena lilacifolia (Pk.) A.H. Smith   Plate 83

PILEUS up to 2 cm in diameter; convex, flattened on top, later centrally depressed; surface viscid; striate; occasionally lavender when young, soon turning dull yellow. FLESH thin; pallid; lacks distinct odor. GILLS decurrent; lilac, becoming pallid. STIPE up to 3 cm long; viscid; lilac to flesh-color. SPORES white; ellipsoid; smooth; nonamyloid; 6-7 x 3-3.5 μ. EDIBILITY unknown.

M. lilacifolia occurs on coniferous logs and old stumps, usually in small clusters, in fall along the Pacific Coast and in summer and fall in the Rocky Mountains. This small fungus, although not common, will attract attention because of its lilac gills.

Mycena occidentalis Murr.

PILEUS up to 2 cm in diameter; broadly conic to campanulate; striate; grayish-brown, lighter near margin. FLESH gray to whitish; lacks distinct odor. GILLS adnate; rather far apart; grayish-white. STIPE up to 5 cm long; smooth except at base, where white hairs are present; gray. SPORES white; ellipsoid; smooth; amyloid; 7-9 x 4-5.5 μ. EDIBILITY unknown.

M. occidentalis is very common on rotting conifer logs in the rainy season occurring from Alaska south to California and Colorado.

M. galericulata (Fr.) S.F. Gray is a species with similar coloring, but with a darker brownish disc. It is much larger, up to 7 cm in diameter, and has a long stipe, sometimes with a pseudorhiza, a rootlike projection for attachment to buried wood.

Mycena pura (Fr.) Quél.

PILEUS up to 4 cm in diameter; broadly convex with umbo to plane or with margin raised; moist; striate; purplish-red to lilac. FLESH moderately thick; purplish or lilac to whitish; odor and taste of radish. GILLS adnate; often intervenose; purplish-gray to whitish. STIPE up to 8 cm in length; relatively thick; similar in color to pileus. SPORES white; ellipsoid; smooth; amyloid; 6-9 x 3-3.5 μ. EDIBILITY unknown.

M. pura grows on humus in coniferous forests and woodland areas and may appear even in grassland. This is one of the most widespread and variable members of the genus *Mycena*. It shows considerable range in size, shape, and color, but the odor of radish when crushed is distinctive.

Mycena purpureofusca Pk.   Plate 84

PILEUS up to 2 cm in diameter; conic to campanulate; moist; striate; dark purple in center to pale lilac near margin. FLESH purplish-gray to whitish; lacks distinctive taste or odor. GILLS adnate; whitish to pale gray, edges dark purple. STIPE up to 8 cm long; slender; smooth above, numerous minute hairs toward base; purplish-gray. SPORES white; ellipsoid; smooth; amyloid; 8-10 x 6-7 μ. EDIBILITY unknown.

M. purpureofusca grows on conifer logs. Its purple gills edges are distinctive.

Mycena rosella (Fr.) Quél.

PILEUS up to 2 cm in diameter; conic to convex; moist; striate; pinkish. FLESH pinkish; lacks distinct odor or taste. GILLS adnate; often connected by veins; whitish to pink. STIPE up to 6 cm long; slender; smooth except for white hairs at base; pinkish-gray. SPORES white; ellipsoid; smooth; amyloid; 7-9 x 4-5 μ. EDIBILITY unknown.

M. rosella sometimes occurs in great numbers on the floors of coniferous forests in autumn in the Pacific Northwest and northern Rocky Mountains. It is a delicate pink mushroom.

Mycena overholtzii A.H. Smith and Solheim

PILEUS 1.5-5 cm in diameter; convex to umbonate in age; smooth; moist; brown or grayish-brown; striate. FLESH soft; white. GILLS adnate; rather far apart; pale gray. STIPE 4-15 cm long, 3-10 mm thick; tapering upward; pinkish-brown; lower half densely covered with white, woolly hairs. SPORES white; elliptical; smooth; 5-7.5 x 3.5-4 μ. EDIBILITY unknown.

M. overholtzii, one of the larger members of the genus, occurs in clusters on conifer logs near melting snowbanks in western mountains. It is unmistakable because of the dense white tomentum on the stipe.

Genus *Marasmius*

Members of *Marasmius* are small tough-fleshed fungi with adnate or adnexed gills and thin, pliant stipes. Spores are white, smooth, and nonamyloid; there is no annulus or volva. Like *Xeromphalina* they are able to revive when moistened. Not much is known about edibility, because the toughness of most species would not make them worthwhile in any event. Only one, *M. oreades*, is of sufficient size and proper texture for cooking, and it is considered to be a choice edible.

Marasmius oreades Fr.

PILEUS 2-6 cm in diameter; convex to campanulate; smooth; dry or moist; reddish-brown in the button stage, becoming light tan or creamy; margin striate and sometimes recurved. FLESH thick; pallid; odor faintly fragrant, taste nutlike. GILLS adnexed, almost free; far apart; pallid. STIPE 4-8 cm long, 3-5 mm thick; equal; tough; dry; pallid above and concolorous with pileus or darker below. SPORES white; ovate or fusiform; smooth; nonamyloid; 7-9 x 4-5.5 μ. EDIBLE and choice; see below.

M. oreades is a common mushroom which grows throughout the temperate zone in grassy places such as lawns, golf courses, and meadows. It often appears in circles known as Fairy Rings (Fig. 25), which become increasingly larger each year as the mycelium grows outward from the starting point. It has a delicate flavor that is said to be enhanced if the fungi are first dried, then placed in water or broth before cooking.

A very poisonous little mushroom, *Clitocybe dealbata* (see below), which is dull grayish-white with decurrent gills, is occasionally found in the same rings.

FIG. 25. *Marasmius oreades* growing in a fairy ring.

Marasmius rotula (Scop. ex Fr.) Fr.

PILEUS 3-10 mm in diameter; convex; umbilicate; radially plicate; white, darker on disc. FLESH membranous; white. GILLS attached to a collar at apex of stipe; far apart; broad; white. STIPE 2-5 cm long; filiform; hollow; smooth; tough; black. SPORES white; fusiform; smooth; 6-9 x 3-4 μ. EDIBILITY unknown.

M. rotula, the Little Wheel fungus, is so named because the gills which are attached to a collar at the top of the stipe resemble the spokes of a wheel. It occurs gregariously on leaves, sticks, and other debris and is easily overlooked because of its short stature.

A very small mushroom that grows in a similar habitat is **M. scorodonius** Fr. The convex to plane, reddish-brown pileus is 5-12 mm in diameter, and the stipe is 2-3 cm long, whitish at apex and blackish or brownish-red below. It has a strong garlic odor when the flesh is crushed, and in Europe is often painstakingly collected for flavoring; it loses potency in cooking and should be added just before serving.

Another common Pacific Coast species that grows on berry canes and the trunks and branches of dead hardwoods is **M. magnisporus** Murr. (also referred to as **M. candidus** Fr.). This umbrella-shaped, semitransparent fungus has veined gills and is white at first, but in age has a pinkish tinge. It is inedible.

Marasmius plicatulus Pk.

PILEUS 1-3.5 cm in diameter; obtusely conic when young, becoming campanulate or convex; pruinose and dry; plicate and striated almost to disc; chestnut-brown. FLESH thin; pliant; pallid. GILLS adnate to nearly free; far apart; flesh-colored to buff or pinkish. STIPE 6-12 cm long, 1.5-3.5 mm thick; brittle; equal; pallid to clay-color, dark chestnut at base; white mycelioid mat. SPORES white; subovoid; non-amyloid; 11-14.8 x 5-6.5 μ. EDIBILITY unknown.

M. plicatulus occurs under both conifers and hardwoods. This is a very attractive mushroom which is called the Pleated Marasmius. Another color form has a dark wine-red pileus and a shiny stipe that is pinkish or red at the apex and blackish below.

Genus *Leucopaxillus*

Because of certain microscopic differences, such as the presence of clamp connections and amyloid spores, the genus *Leucopaxillus* was erected from some species of such genera as *Pleurotus*, *Clitocybe*, and *Tricholoma*. The macroscopic feature which makes members of *Leucopaxillus* immediately recognizable is the abundant white mycelium which is matted in the duff around the base of most species. The spores are white, and the gills are decurrent to sinuate. *Leucopaxillus* grows on the ground around conifers and hardwoods. Although none is known to be poisonous, several species have an unpleasant odor or a bitter taste.

Leucopaxillus amarus (Alb. and Schw.) Kühn.

PILEUS 4-12 cm in diameter; convex with an inrolled margin; smooth; dry; chocolate-brown or reddish-brown, becoming paler with age. FLESH firm; white; taste bitter. GILLS adnate; close; white. STIPE 4-6 cm long, 8-45 mm thick; equal to bulbous; white; white mycelium at base. SPORES white; subglobose; amyloid; 4.3-6 x 3.7-5 μ. IN-EDIBLE.

L. amarus is a very common species under conifers or oaks in most of the western states. It may occur singly or scattered.

Leucopaxillus albissimus (Pk.) Sing.

PILEUS 6-20 cm in diameter; convex to plane or depressed; smooth; dry; white. FLESH soft; white. GILLS decurrent; crowded; white. STIPE 8-15 cm long, 1.5-3 cm thick; equal or bulbous; white; white mycelium at base. SPORES white; elliptical; amyloid; 5.5-8.5 x 4-5.5 μ. NONPOISONOUS, but bitter.

L. albissimus is of fairly widespread occurrence in areas where oaks or conifers grow in the West.

Genus *Clitocybe*

Members of *Clitocybe* are small to medium-sized mushrooms with decurrent gills, a fibrous stipe which does not separate easily from the pileus, and white, yellow, or pinkish-buff spores. There is no annulus or volva.

There are some edible species, but none that is outstanding, and several are mildly to seriously poisonous. The most toxic species is *C. dealbata* because of its high muscarine content. It is particularly dangerous to children because it often occurs in lawns where they play.

KEY TO SPECIES OF *CLITOCYBE*

1a. With strong, fragrant odor 2
 b. Not as above 3
  2a. Pileus and gills dull
   bluish-green . . . . *C. odora*, p. 243
   b. Pileus pale pinkish-orange . . . . . . *C. olida*, p. 244
   c. Pileus grayish-brown; stipe with large
    basal bulb . . . . *C. clavipes*, p. 246
3a. Pileus pinkish-tan; funnel-shaped . . . *C. gibba*, p. 244
 b. Not as above 4
  4a. Fruiting body small; all parts
   grayish-white . . . . *C. dealbata*, p. 244
   b. Not as above 5
5a. Pileus pale orange or yellowish-orange 6
 b. Pileus grayish, brownish-gray, or blackish-brown . . 7
  6a. Gills pale orange,
   far apart . . . . *C. luteicolor*, p. 245
   b. Gills bright orange,
    crowded . . . . *C. aurantiaca*, p. 245
7a. Pileus gray to nearly pallid; gills crowded,
   adnate to short-decurrent; strong,
   unpleasant odor . . . . *C. nebularis*, p. 246
 b. Pileus blackish-brown (in youth) to
   brownish-gray 8
  8a. Stipe smooth;
   gills close . . . . *C. avellaneialba*, p. 245
   b. Stipe scurfy; gills far apart . . . *C. atrialba*, p. 246

Clitocybe odora (Fr.) Kumm.   Plate 85

PILEUS 3-10 cm in diameter; convex to expanded; smooth; dry; dull bluish-green or grayish-green. FLESH thin; greenish-white. GILLS adnate to short-decurrent; close; bluish-green. STIPE 2-9 cm long, 4-16 mm thick; equal or slightly enlarged at base; whitish, buff, or concolorous with pileus. SPORES pinkish-buff; elliptical; smooth; 5-9 x 3.5-5 μ. EDIBLE, but poor.

C. odora often appears in abundance under conifers and is fairly common in the West. Its anise odor and green coloration make it easily identifiable.

A much rarer species, **C. olida** Quél., has a pale pinkish-orange pileus and a sweet, fruity fragrance. Its edibility is unknown.

Clitocybe gibba (Fr.) Kumm.
Clitocybe infundibuliformis (Fr.) Quél.

PILEUS 3-9 cm in diameter; plane or depressed at first, becoming deeply funnel-shaped (infundibuliform); smooth; moist; pinkish-tan to flesh-color. FLESH thin; white. GILLS decurrent; close; forked; whitish to buff. STIPE 3-7 cm long, 4-12 mm thick; equal or slightly enlarged at base; smooth; dry; whitish; dense white mycelioid mat at base. SPORES white; elliptical; smooth; 5-10 x 3.5-5.5 μ. EDIBLE.

C. gibba, the Funnel-Shaped Clitocybe, is common and appears singly or gregariously in conifer and hardwood forests. It is of widespread occurrence ranging north to Alaska.

Clitocybe dealbata (Sow. ex Fr.) Kumm.

PILEUS 1.5-5 cm in diameter; convex to plane, becoming depressed; smooth; dry; grayish-white. FLESH thin; whitish. GILLS decurrent; close; narrow; grayish-white. STIPE 1-5 cm long, 2-6 mm thick; tough; smooth; equal or tapering downward; grayish-white. SPORES white; elliptical; smooth; 4-5 x 2.5-3 μ. POISONOUS.

C. dealbata is widely distributed. It occurs in open woods or grassy places such as lawns, often among edible mushrooms such as *Marasmius oreades* (see above). Since it has a high muscarine content, it is particularly hazardous to small children.

Clitocybe luteicolor (Murr.) Bigelow and Smith

PILEUS 1.5-3 cm in diameter; convex, becoming depressed; sometimes with slightly raised margin; fibrous; pale orange or yellowish-orange. FLESH thin; brittle; concolorous with pileus. GILLS decurrent; far apart; pale orange, appearing waxy. STIPE 3-5 cm long; 3-5 mm thick; smooth; equal or tapering downward; orange. SPORES white; ellipsoid; smooth; nonamyloid; 8-10 x 4-5 μ. EDIBILITY unknown.

C. luteicolor is conspicuous on conifer logs, where this colorful little fungus occurs singly or gregariously.

Clitocybe aurantiaca (Fr.) Studer
Hygrophoropsis aurantiaca (Wulfen ex Fr.) Maire

PILEUS 2.5-6 cm in diameter; convex to plane or depressed; subtomentose; orange-yellow to brownish-orange. FLESH thin at margin; soft; whitish to pale orange. GILLS decurrent; close; thin; forked; bright orange to salmon. STIPE 2.5-9 cm long; 5-12 mm thick; tapering upward; concolorous with pileus. SPORES white; elliptical; smooth; 5-7 x 3-4 μ. POISONOUS; see below.

C. aurantiaca occurs on the ground or rotten logs in conifer forests and is widely distributed. It also has a form with a brown pileus and stipe, but that is much less common than the orange form. There are conflicting statements about its edibility, and perhaps it is a matter of personal sensitivity. However, it is best to consider it poisonous, since it has been known to cause illness.

Clitocybe avellaneialba Murr.

PILEUS 8-12 cm in diameter; convex; sometimes bluntly umbonate; becoming plane or depressed; fibrillose; moist; blackish-brown, becoming paler in age. FLESH thin; white; odor and taste unpleasant. GILLS decurrent; close; pallid to cream-color. STIPE 6-12 cm long; 1-3 cm thick; stuffed or hollow; smooth; decidedly clavate, up to 4 cm at base;

blackish-brown to gray. SPORES white; fusiform; smooth; nonamyloid; 8-10 x 4-5.5 μ. EDIBILITY unknown.

C. avellaneialba, a large and striking mushroom with its contrasting dark and light coloration, occurs on humus or rotting wood under conifers or alders along the Pacific Coast.

A very similarly shaped and colored mushroom, **C. clavipes** (Fr.) Kumm., is edible. It has an abrupt basal bulb and fragrant odor, and its spores are white, ovoid, smooth, 6-10 x 3.5-5 μ.

Another smoky or blackish-brown mushroom, **C. atrialba** Murr., has a densely scurfy stipe that is often enlarged at the apex; otherwise it resembles *C. avellaneialba*. Its spores are white, subellipsoid to ovoid, amyloid; 10-12.5 x 7-9 μ. Nothing is known about its edibility.

Clitocybe nebularis (Fr.) Kummer

PILEUS 6-15 cm in diameter; convex to plane or depressed; dry; smooth; pruinose; margin inrolled; pale brownish to gray or nearly pallid. FLESH tough; thick; white; odor and taste strong and unpleasant. GILLS adnate to decurrent; close; whitish. STIPE 8-10 cm long, 2.5-4 cm thick; often clavate; shiny white. SPORES pale yellow; elliptical; smooth; 5.5-8 x 3.5-4.5 μ. POISONOUS; see below.

C. nebularis, the Cloudy Clitocybe, is a common species that grows on the ground under conifers in western United States and Canada. Although it is listed as edible in some works, it has been known to cause gastrointestinal disturbance. Its disagreeable odor and taste would probably discourage the mushroom hunter in any case. It is sometimes parasitized by *Volvariella surrecta* (see above).

Genus *Laccaria*

Laccaria contains few species, and these are readily recognizable by their thick, rather widely spaced, purple to flesh-colored gills, which are adnate to slightly decurrent, and the lack of either annulus or volva. Spores are white and, in most species, ornamented. Two of the most common species found in the West are *L. laccata* and *L. amethystina*.

Laccaria laccata (Fr.) Berk. and Br.   Plate 86

PILEUS up to 8 cm in diameter; convex to plane, with center slightly depressed; dry to moist; sometimes minutely scurfy; flesh-colored to pinkish-brown. FLESH thin; pallid to flesh-color; odor and taste not distinctive. GILLS rather far apart; sometimes slightly decurrent; flesh-color. STIPE up to 10 cm long; equal; fibrous; surface slightly scaly; concolorous with pileus. SPORES white; subglobose; minutely ornamented; 7.5-10 x 7-8.5 μ. NONPOISONOUS, but not recommended.

L. laccata is a very common, widespread, and also very variable species. It is terrestrial on humus, growing in spring, summer, and fall.

L. amethystina (Bolt. ex Hook.) Murr. is quite similar, but the entire fruiting body is a deep violet. It is not as common as *L. laccata*, but occurs in the same habitat; both tend to be gregarious. **L. ochropurpurea** (Berk.) Pk. also has intensely violet gills, but the pileus and stipe are yellowish to grayish-tan. Members of the genus are nonpoisonous but they are not very palatable.

Genus *Lyophyllum*

Lyophyllum is difficult to distinguish from such white-spored genera as *Tricholoma*, *Collybia*, and *Clitocybe* on the basis of macroscopic characters. Microscopically, the genus is differentiated by the presence of numerous granules in the basidia which darken when treated with a carmine stain. Most species of *Lyophyllum* are various shades of gray or brown, and they are terrestrial.

Lyophyllum montanum A.H. Smith   Plate 87

PILEUS 2-7 cm in diameter; margin inrolled at first but soon becoming plane; smooth; grayish-brown with a pale gray bloom which washes or rubs off. FLESH moderately thin; pale gray; odor and taste slightly farinaceous. GILLS adnexed; thin; close; pale gray to grayish-tan; several tiers of lamellulae. STIPE 4-7 cm in length, somewhat enlarged basally and surrounded by white mycelioid mat; concolorous with pileus; hollow; cartilaginous. SPORES white; ellipsoid to cylindrical; smooth; 10-11 x 3.5-4.7 μ. EDIBILITY unknown.

L. montanum is one of the most common species of fungi found in summer on the edge of snowbanks in the mountains of the West. This beautiful mushroom has a pale gray bloom on the surface of the pileus which almost blends with the color of the edge of the snowbank through which it may be growing. It occurs singly, or several fruiting bodies may grow in close approximation to one another.

Lyophyllum decastes (Fr.) Sing.
Clitocybe multiceps (Pk.)

PILEUS 6-14 cm in diameter; broadly convex with margin inrolled when young; smooth; gray to grayish-brown; moist. FLESH white and firm. GILLS adnate; white. STIPE up to 10 cm long; thick; white. SPORES white; broadly elliptical; smooth; 5-7 x 5-6 μ. EDIBILITY questionable.

L. decastes is a widely distributed though rarely abundant mushroom that usually occurs in dense clusters on the ground. Although listed as edible in some books, we know of a few cases of illness caused by it.

Genus *Collybia*

Collybia is a genus with a large number of species, although there have been several transfers to other genera such as *Lyophyllum*, *Flammulina*, and *Tricholomopsis*. Its members have a convex pileus with inrolled margin when young; adnate or adnexed gills; a firm and brittle stipe; and smooth, white, pale cream or buff spores; there is no annulus or volva. Its habitat may be the ground, decayed wood, or forest debris. Several are listed as edible, but none is outstanding.

Collybia dryophila (Bull. ex Fr.) Kumm.

PILEUS 2.5-6 cm in diameter; convex to plane, sometimes slightly umbonate; rusty-brown, reddish-brown, or yellow-brown; margin often raised and wavy. FLESH thin; watery; white to pallid. GILLS adnate to adnexed; crowded; narrow; pallid. STIPE 4-7 cm long, 2-6 mm thick; equal; smooth; pallid or brownish-tinged; many white rhizomorphs. SPORES white; elliptical; smooth; 5-7 x 3-3.5 μ. EDIBILITY questionable.

C. dryophila is called the Oak-loving Collybia because it occurs in hardwood forests or mixed hardwood and conifer woodlands. It is common throughout North America and abundant on the Pacific Coast. It was formerly listed as edible, but Miller (1972) cites cases of severe gastrointestinal illness caused by it; we have eaten it sparingly with no trouble.

Collybia butyracea Fr.

PILEUS 2-8 cm in diameter; convex and umbonate; smooth; slippery feel; reddish-brown. FLESH soft; thick; whitish. GILLS adnexed, nearly free; crowded; white, stains brownish; edges crenulate. STIPE 3-7 cm long, 4-12 mm thick; tapering upward; striate; paler than pileus. SPORES pale buff; elliptical; smooth; 5-7 x 3-3.4 μ. EDIBLE.

C. butyracea, the Buttery Collybia, is common and widely distributed. It is similar to *C. dryophila* (see above) in appearance and habitat, but it can be distinguished by the slippery feel and buff spores. It is said to have an equally good flavor.

Collybia familia Pk.   Plate 88

PILEUS 1-3.5 cm in diameter; convex; smooth; hygrophanous; whitish or buff to pale brown. FLESH soft; thin;
concolorous with pileus. GILLS adnexed; crowded; whitish.
STIPE 4-8 cm long, 2-3 mm thick; stuffed, then hollow;
equal; smooth; joined with others at base with mycelioid
tomentum. SPORES white; subglobose; smooth; 3-4 x 3 μ.
EDIBILITY unknown.

C. *familia* is widely distributed and appears in dense
clusters on decaying conifer logs.

Another caespitose species, **C. acervata** (Fr.) Kumm., is
vinaceous-brown. It is not poisonous, but has a very bitter
taste, and is also widely distributed.

Collybia umbonata Pk.   Plate 89

PILEUS 3-7 cm in diameter; bluntly conic; becoming expanded; smooth; chestnut-brown or yellowish-brown. FLESH
thick; firm; pallid. GILLS adnate; yellowish or pallid.
STIPE 8-14 cm long, 1-2 cm thick; tapering to a long taproot;
yellowish-brown. SPORES white; ellipsoid; smooth; 5-6 x 3
μ. EDIBILITY unknown.

C. *umbonata* occurs only in northern California, during
the fall and winter, always in association with Coast
Redwoods.

Genus *Schizophyllum*

Schizophyllum means "split leaf" and describes the unique
character of the gills in the genus: they are longitudinally
split about halfway down from the edge and are recurved.

Schizophyllum commune Fr.

PILEUS 1-3 cm in diameter; sessile; fan-shaped; thin; extremely hairy; grayish-buff when moist, white when dry.
FLESH thin; tough; gray. GILLS radiate from the point of

attachment on wood; split; hairy; concolorous with pileus or pinkish. SPORES white; cylindrical; 3-4 x 1-1.5 μ. IN-EDIBLE.

S. commune is cosmopolitan and the only species in the genus in the temperate zone. It occurs on wood and can severely damage structures such as telephone poles, and it is also destructive to fruit trees and berry canes.

Genus *Panus*

Panus differs from a related genus, *Panellus* (see below) by having nonamyloid spores and very tough flesh. The gills are decurrent, and the stipe may be lateral, eccentric, or lacking. Spores are white or buff. Members of *Panus* are lignicolous.

Panus conchatus (Bull. ex Fr.) Fr.

PILEUS 5-15 cm in diameter; broadly convex to plane or depressed; tan, lilac, or reddish-brown; margin inrolled and often wavy. FLESH tough; pallid. GILLS decurrent; narrow; often forked; at first pallid, becoming tan or reddish-violet. STIPE lateral or eccentric; 2-3 cm long, 1-3 cm thick; tough; concolorous with pileus, but covered with violet hairs. SPORES white; elliptical; smooth; 5-7.5 x 2.5-3 μ. INEDIBLE.

P. conchatus occurs on hardwood stumps or logs, usually in caespitose clusters, but it is rather rare; we have found some very large specimens in Oregon.

Genus *Phyllotopsis*

Phyllotopsis contains only one species, which was formerly classified as *Pleurotus* although its pink spores did not conform with that white- or lilac-spored genus.

Phyllotopsis nidulans (Pers. ex Fr.) Sing.

PILEUS 3-8 cm in diameter; sessile; kidney-shaped to circular; coarsely hairy; pale orange to buff; margin inrolled. FLESH tough; layer above orange-buff, paler layer below; odor usually strong and unpleasant. GILLS adnate; close; bright orange-buff. SPORES salmon-pink; elliptical to allantoid; smooth; 5-7 x 2-2.5 μ. INEDIBLE.

P. nidulans is widely distributed and common, occuring in clusters on both conifers and hardwoods or logs, and is quite destructive to a living host. It is characterized by a densely tomentose orange pileus, adnate gills, lack of stipe, and salmon-pink spores, which will fade to white after a while in the herbarium.

Genus *Pleurotus*

Members of *Pleurotus* occur on trees, decaying logs, or buried wood. They may be sessile or have a lateral or eccentric stipe, and the gills are adnexed to decurrent, sometimes anastomosing at the point of attachment. The spores are white or lilac, smooth, and nonamyloid. One of the choicest and most popular edible species is *P. ostreatus* which appears in overlapping clusters. No species of *Pleurotus* is known to be poisonous.

Pleurotus ostreatus Fr.

PILEUS 2-30 cm in diameter; fan-shaped to convex; white to gray or dull brown; margin sometimes wavy. FLESH thick; firm; white. GILLS decurrent, often anastomosing; white or grayish. STIPE lacking, or lateral or eccentric; dry; equal; white. SPORES white; elliptical; smooth; 8-10 x 3.5-4.5 μ. EDIBLE and choice.

P. ostreatus is widely distributed and is called the Oyster Mushroom. This name comes from the shell-shaped pileus and does not refer to the flavor, though it is highly regarded as an edible.

A closely related species, **P. sapidus** (Schulzer) Kalch., has lilac spores, and is equally edible. It, too, is widely distributed.

Pleurotus porrigens (Fr.) Gill.   Plate 90
Pleurocybella porrigens (Pers. ex Fr.) Sing.

PILEUS 4-7 cm in diameter; fan-shaped; dry; minutely tomentose; white; margin inrolled at first. FLESH thin; pliant; white. GILLS extend to base; close; white; sessile. SPORES white; subglobose; smooth; nonamyloid; 6-7 x 5-6.5 μ. EDIBLE.

P. porrigens occurs in dense clusters on conifer logs. These beautiful, pure white mushrooms are commonly called Angel's Wings.

Genus *Panellus*

Panellus includes certain species of lignicolous fungi that have white to yellow, amyloid spores and gills with even edges. Because of the amyloid spores, *Panellus* has been separated from a closely related genus, *Panus* (see above). *Panellus* may be stipitate or sessile and, if a stipe is present, the gills are attached.

Panellus serotinus (Fr.) Kummer
Pleurotus serotinus Fr.

PILEUS up to 10 cm in diameter; fan-shaped; smooth; viscid; greenish. FLESH white; firm; lacks distinct odor or taste. GILLS adnate; yellowish-tan to pale orange. STIPE lateral, very short, 5-20 mm long. SPORES white; narrowly oblong; smooth; 4-5.5 x 1-1.5 μ. EDIBLE, but poor.

P. serotinus is a firm, rather tough fungus that grows on logs or stumps in a series of shelves, or it may be solitary. It is found in the fall in both coniferous and nonconiferous forests and is widely distributed. Although edible, it is said to be of poor quality.

P. stypticus (Bull. ex Fr.) Karst. is a poisonous species, but its strong, astringent taste would seem to deter anyone from eating it. Its small, somewhat kidney-shaped pileus is pale tan or cinnamon and scurfy. The gills are cinnamon and end abruptly at the very short lateral stipe. It is common on logs and stumps and is widely distributed.

Genus *Lentinus*

Lentinus contains fleshy though tough fungi whose spores are white to cream-color and nonamyloid. The edges of the gills are minutely toothed or irregularly notched.

Lentinus lepideus Fr.

PILEUS up to 30 cm in diameter in large fruiting bodies, but frequently smaller; convex; surface dry; tan to creamy-white with numerous large brown patches. FLESH white; thick; tough. GILLS adnexed to decurrent on stipe; whitish when young, becoming cream to tan with pinkish tinge; edges coarsely serrate; moderately far apart; numerous lamellulae. STIPE up to 8 cm in length; thick, tapering downward; whitish at first, becoming reddish-brown; superior membranous annulus present when young. SPORES white; smooth; cylindric; 12-15 x 5-6 μ. EDIBLE when young.

L. lepideus is found on conifer logs and stumps in summer in the western mountains. The large, tough, fruiting body with brown patches on the pileus, and the serrate edges of the gills, provide good field characters. *L. lepideus* may cause considerable damage to timbers supporting manmade structures and, because it sometimes infests railroad ties, it is called the Train Wrecker. Spores of species in the West measure larger than those from other areas.

Genus *Xeromphalina*

Members of *Xeromphalina* are very small fungi which occur in numbers on wood or humus. Like *Marasmius*, they revive when moistened, but they are differentiated by their amyloid spores. The gills are adnate to decurrent, the stipe is brown or blackish and pliant, and the spores are white.

Xeromphalina campanella (Fr.) Kühn. and Maire

PILEUS 3-25 mm in diameter; convex to broadly convex with a central depression; moist; smooth; dull yellowish-orange to orange-brown; margin at first inrolled, striate. FLESH thin; pliant; yellowish. GILLS decurrent; far apart; intervenose; yellowish or orange. STIPE 1-4 cm long; very thin; tough and pliant; hollow; yellowish at apex, dark reddish-brown below with yellowish-brown hairs at base. SPORES white; elliptical; smooth; amyloid; 6-7.5 x 2.5-3.5 μ. EDIBILITY unknown.

X. campanella occurs in dense caespitose clumps on decaying conifer wood. This cosmpolitan species is quite common.

Another common species, **X. cauticinalis** (Fr.) Kühn. and Maire, is found on needles, sticks, or leaves and is widely distributed. It has a deep yellow pileus and a much longer stipe, up to 8 cm long. Its edibility is unknown.

Genus *Tricholomopsis*

In 1939, species transferred from *Tricholoma*, *Collybia*, *Clitocybe*, and *Melanoleuca* were classified as *Tricholomopsis*. Its members occur on wood, the pileus is usually fibrillose, and the gills are frequently yellow. They may be adnate, adnexed, sinuate, or decurrent, and there are prominent cheilocystidia on the edges. The stipe is central, and there is no annulus or volva. The spores are white, ellipsoid to globose, smooth, and nonamyloid.

Tricholomopsis platyphylla (Pers. ex Fr.) Sing.

Collybia platyphylla (Pers. ex Fr.) Moser

PILEUS 5-12 cm in diameter; convex to plane, sometimes with low umbo or depressed; surface dry or moist; fibrillose; blackish-brown to gray. FLESH watery; gray. GILLS adnate to adnexed; rather far apart; broad; white to grayish. STIPE 6-12 cm long, 1-2 cm thick; equal or slightly enlarged at base; stuffed, becoming hollow; white to grayish. SPORES white; ellipsoid; smooth; nonamyloid; 7-10 x 4.5-6 μ. EDIBILITY unknown.

T. platyphylla occurs on logs or stumps of both hard-woods and conifers. It is widely distributed and common.

Tricholomopsis decora (Fr.) Sing.

PILEUS up to 6 cm in diameter; convex when young, becoming plane and centrally depressed when mature; surface moist; yellow with minute gray to brown scales. FLESH rather thin; yellow; lacks any distinct odor or taste. GILLS adnexed to adnate or decurrent with a tooth; yellow. STIPE up to 8 cm long; equal; yellow. SPORES white; ovoid; smooth; 6-7.5 x 4-5 μ. EDIBILITY unknown.

T. decora grows on rotten wood in the forest. Its resemblance to *T. rutilans* (see below) is easily seen in the field, but *T. decora* lacks the purplish-red color seen on both pileus and stipe of *T. rutilans*.

Tricholomopsis rutilans (Fr.) Sing.

PILEUS up to 12 cm in diameter; hemispheric with incurved margin when young, becoming nearly plane with age; yellow with numerous purplish-red scales forming solid mat toward center of pileus. FLESH thick; yellow. GILLS adnate to adnexed; close; edges crenulate; yellow. STIPE up to 12 cm or more in length; 1-3 cm in diameter; becoming hollow in age; yellow at apex but covered with numerous purplish-red scales for most of length. SPORES white; ovoid to near globose; smooth; 5-6 x 4-5 μ. EDIBLE.

T. rutilans grows on wood, usually in caespitose clumps. It is associated with coniferous forests from Alaska to California and throughout the Rocky Mountains during the rainy season. This is one of our most beautiful gill fungi because of its large size, purplish-red and yellow colors, and cleancut appearance.

Genus *Flammulina*

Members of *Flammulina* were transferred from the genus *Collybia*. *Flammulina* is characterized by attached gills and white, nonamyloid spores. There is no annulus or volva. Only one species is common in North America.

Flammulina velutipes (Fr.) Sing.   Plate 91
Collybia velutipes Fr.

PILEUS 2-10 cm in diameter; convex to nearly plane in age; viscid; yellow to reddish-yellow or yellowish-brown; margin often irregularly elevated. FLESH thick; white or yellowish. GILLS adnexed; broad; rather far apart; creamy or yellowish; edges minutely hairy. STIPE 2-7 cm long, 3-8 mm thick; firm; hollow in age; yellow at apex, densely covered with velvety, brown or blackish-brown hairs. SPORES white; elliptical; smooth; nonamyloid; 7-9 x 3-4 μ. EDIBLE and choice; see below.

F. velutipes, the Velvet-Footed Mushroom, is a common species in western North America from Alaska south. It fruits in cold weather on both living and dead wood and is considered a very fine edible. The viscid pellicle should be removed and the tough stipe discarded before cooking.

Genus *Omphalotus*

Until recently **O. olearius** was considered to be the only species in *Omphalotus*, but in 1976 Bigelow, Miller and Thiers determined that our western fungus is a new species, *O. olivascens*. Its distribution is uncertain at the present time, but it is fairly common in north-central California. Macroscopically it differs from *O. olearius* mainly by duller orange coloring with distinctive olive overtones, and there were also differences observed in cultures of the two species. *O. olearius* is presently known only from the East.

Omphalotus olivascens Bigelow, Miller, and Thiers

PILEUS 4.5-25 cm in diameter; convex, becoming broadly convex to plane, centrally depressed in larger pilei; margin at first inrolled, becoming wavy or lobed in age; dull orange, disc often olive-tinged. FLESH watery; tough; dull orange, becoming olivaceous in age. GILLS decurrent; sometimes forked and intervenose; olive spotted with dark yellow when young, becoming oranger in age; luminescent. STIPE 4-22 cm long, 7-80 mm thick, tapering downward; solid; olive to olivaceous-yellow; stained rusty-brown in age; annulus and volva lacking. SPORES cream-color; globose to broadly elliptical; smooth; nonamyloid; 7-8 x 6-6.5 μ. POISONOUS.

O. olivascens generally grows in large clusters at the base of stumps, especially those of oak, or attached to buried roots. The Toxicology Committee of the Mycological Society of San Francisco was given a first-hand report of poisoning from two men who had collected it in March, 1977, growing at the base of a dead oak. (They ate it because they thought it was a chanterelle!) One hour after ingestion they developed a severe gastrointestinal disturbance that lasted ten hours, and one man was hospitalized. Both fully recovered after dehydration was corrected.

Genus *Lepista*

The species of *Lepista* mentioned here were transferred from the white-spored genus *Tricholoma* because they have pinkish or flesh-colored spores. Some authorities classify them as *Clitocybe*.

Lepista nuda (Bull. ex Fr.) Cke.   Plate 92
Tricholoma personatum (Fr.) Kumm.
Clitocybe nuda (Fr.) Bigelow and Smith

PILEUS 4-15 cm in diameter; convex to plane; smooth; moist; pale bluish-gray to various shades of violet, buff, or lavender, disc brownish in age; margin inrolled at first, often wavy in age. FLESH thin; soft; lavender or pale lilac-buff; odor faintly fragrant. GILLS adnexed or sinuate; broad; close; violet or pale lilac. STIPE 3-7 cm long, 1-2.5 cm thick; equal or bulbous at base; pale violet. SPORES dingy pink or flesh-colored; elliptical; minutely roughened; 5-5.8 x 3.5-5 μ. EDIBLE and choice.

L. nuda occurs commonly throughout the temperate zone under hardwoods or conifers, either singly or gregariously, and often in fairy rings.

L. irina (Fr.) Bigelow and Smith has a similar stature but lacks the violet tones of *L. nuda*. It is whitish to pallid or pale tan with white gills which become dingy buff. The spores are pale pinkish-buff and measure 7-10 x 4-5 μ. Though it is also listed as choice, it is known to have caused gastrointestinal disturbance in some people. It also has wide distribution.

Genus *Tricholoma*

Tricholoma is well represented in most forested areas of the West. The various species are fleshy and mostly of medium to large size. They usually grow on the ground, only rarely on wood. The gills are attached to the central stipe, are either sinuate or adnexed, and generally lack cystidia; the stipe is fleshy or fibrous. Neither an annulus nor a volva is present. Spores are white to very pale cream-color and may be subglobose, ovoid, ellipsoid, fusiform, or angular in shape, and nonamyloid. Several species are edible and of mild flavor, but many are suspected or known to be poisonous.

KEY TO SPECIES OF *TRICHOLOMA*

Tricholoma flavovirens (Fr.) Lund.
Tricholoma equestre (Fr.) Kumm.

PILEUS 6-12 cm in diameter; broadly convex with margin often irregularly wavy; slightly viscid; yellow but becoming brownish near center, sometimes with radial streaking. FLESH white to pale yellow; odorless. GILLS adnexed to sinuate; yellow; easily broken. STIPE up to 20 cm in length but usually less than 10 cm; equal; up to 2.5 cm in diameter; smooth on surface; yellow externally; white within. SPORES white; ellipsoid; smooth; 6-7.5 x 4-5 μ. EDIBLE.

T. flavovirens is a fairly common species in coniferous woods along the Pacific Coast in fall and winter and in the Rocky Mountains in summer and fall. Its bright yellow pileus and large size make it a conspicuous part of the fungus flora.

Growing in the same habitat in parts of the Pacific Northwest is a closely related species, **T. sejunctum** (Sow. ex Fr.) Quél. This is slightly smaller than *T. flavovirens* and has paler gills, which may be white in some fruiting bodies. Furthermore, the pileus is campanulate or broadly conic and much of the yellow is obscured by sooty lines radiating from the dark disc. Its edibility is unknown.

T. sulphureum (Bull. ex Fr.) Kumm. resembles *T. flavovirens* in color but is not viscid and has a disagreeable odor; it is poisonous.

Tricholoma imbricatum (Fr.) Quél.

PILEUS 5-10 cm in diameter; convex when young with margin inrolled, becoming campanulate, often with a distinct umbo, with age; surface dry and minutely fibrillose; reddish-brown. FLESH thick, firm; white but becomes reddish when cut; odor and taste not distinctive. GILLS adnexed; whitish but soon staining reddish-brown. STIPE 4-10 cm long and 2 cm thick; reddish-brown. SPORES white; ellipsoid; smooth; 5.5-7 x 4-5.5 μ. EDIBLE.

T. imbricatum is often gregarious and is fairly common in coniferous forests of the Pacific Coast in fall and winter and the Rocky Mountains in summer and fall.

Tricholoma pardinum Quél.   Plate 93

PILEUS 6-15 cm in diameter; convex to nearly plane; surface dry, white with numerous minute, dark blackish-gray or brownish-gray scales, less noticeable toward margin. FLESH thick; white to grayish. GILLS sinuate; whitish. STIPE 4-10 cm long, moderately thick, slightly enlarged at base; smooth; whitish. SPORES white; ellipsoid; smooth; 8-10 x 5.9-7 μ. POISONOUS.

T. pardinum is common in the coniferous forests of the Rocky Mountains in fall and along the Pacific Coast in fall and winter. Its large size and minutely scaly, blackish-gray, unstreaked pileus help to identify this poisonous species.

Tricholoma saponaceum (Fr.) Kumm.

PILEUS 2-8 cm in diameter; convex; olive-green, becoming darker on disc; smooth. FLESH thick; white, turning pink when cut; soapy in taste and odor. GILLS adnate or with a decurrent tooth; white, often with a dingy green tinge. STIPE 6-10 cm long, 2 cm thick; white but easily stains pink; solid. SPORES white; ellipsoid; smooth; 5.5-7 x 3.5-5 μ. INEDIBLE.

T. saponaceum occurs in all of the West, but is most common in parts of the Rocky Mountains; we have found it very abundant in Yellowstone National Park in summer in Ponderosa Pine forests. The olive-gray pileus, soapy odor, and tendency of the flesh to stain pink when cut or bruised make for easy identification in the field.

Tricholoma terreum (Schaeff. ex Fr.) Kumm.
Plate 94

PILEUS 3-7 cm in diameter; convex to plane, sometimes with a small umbo; surface dry and minutely scaly or woolly; gray except near margin, where whitish. FLESH thin; white to pale gray; no distinctive odor or taste. GILLS adnexed; whitish when young, becoming pale gray with age. STIPE 2.5-5 cm long and relatively slender; whitish to pale gray. SPORES white; ovoid; smooth; 6-8 x 3.5-5.5 μ. NOT RECOMMENDED; see below.

T. terreum is widespread and common in most forested areas in fall and winter. It is the smallest *Tricholoma* described here. In color it somewhat resembles *T. pardinum* (see above), but the small size and minutely floccose surface of the pileus, as well as the rather distinct white margin, serve as good field characters for identification. Although not regarded as poisonous, it should be avoided because of the possibility of confusing it with the larger *T. pardinum* which is poisonous.

Tricholoma vaccinum (Pers. ex Fr.) Kumm.
Plate 95

PILEUS up to 8 cm in diameter; convex when young, becoming more or less campanulate with age; dry; surface has conspicuous reddish-brown scales, whitish flesh exposed between; disc solid reddish-brown. FLESH firm; whitish but stains reddish when cut; taste unpleasant. GILLS adnexed to sinuate; whitish, staining reddish-brown. STIPE up to 8 cm long, somewhat enlarged basally; reddish-brown with fibrillose scales; hollow. SPORES white; ovoid; smooth; 5-7 x 4.5 μ. EDIBILITY unknown.

T. vaccinum resembles *T. imbricatum* (see above) in many respects and often grows in the same general area. The surface of the pileus, however, is marked by large concentric rows of reddish-brown scales.

Tricholoma virgatum (Fr.) Gill.   Plate 96

PILEUS up to 10 cm in diameter; conic, becoming nearly plane with age, often with a slight umbo; dry; slightly fibrillose; light grayish-white near margin, becoming darker bluish-gray toward disc; radially streaked with darker gray. FLESH thick; white to grayish; taste acrid. GILLS adnexed; white when young, becoming gray with age. STIPE up to 12 cm in length, moderately thick; equal; minutely fibrillose; whitish. SPORES white; ellipsoid; smooth; 6-7.5 x 5-6 μ. EDIBILITY questionable.

T. virgatum is common in coniferous forests during the rainy season along the Pacific Coast and in the Rocky Mountains. It is a very attractive-looking fungus with its streaked gray pileus and large size.

GLOSSARY OF
MYCOLOGICAL TERMS

ACICULAR: needle-shaped.

ACUTE: pointed.

ADNATE: united, such as gills or tubes that are broadly attached to the stipe.

ADNEXED; describing gills notched next to the stipe.

AGARIC: a gilled mushroom.

ALLANTOID: sausage-shaped.

AMYLOID: a bluish-gray or violet reaction (of spores) in Melzer's solution.

ANASTOMOSE: to join together by cross-connections.

ANNULUS: a membranous ring on the stipe.

APEX: the top.

APICULUS: a short projection on the end of a spore.

APOTHECIUM: the cup-shaped or urn-shaped fruiting body in the Ascomycetes.

APPENDICULATE: decorated with hanging fragments, as on the margin of the pileus.

APPRESSED: closely pressed against a surface.

AREOLATE: describing a surface cracked into small areas.

ASCUS (pl. ASCI): a tubelike sac in which spores develop in the Ascomycetes.

AXIL: an angle, as between branches in coral fungi.

BASIDIOLE: a small immature basidium lacking sterigmata.

BASIDIUM (pl. BASIDIA): a clublike structure on which spores are borne in the Basidiomycetes.

BOLETE: the common name for a member of the family Boletaceae.

CAESPITOSE: clumped or clustered.

CAMPANULATE: bell-shaped.

CAPILLITIUM: the mass of sterile hyphae within a puffball.

CAPITATE: having a head.

CARPOPHORE: the fruiting body of a mushroom, its spore-producing organ.

CHARACTER: a feature by which a mushroom may be identified.

CHEILOCYSTIDIA: sterile cells on the edge of a gill.

CHLOROPHYLL: the green coloring matter, lacking in fungi, which enables most plants to manufacture their food from carbon dioxide and water in the presence of sunlight.

CLAVATE: club-shaped.

CONCOLOROUS (with): of the same color (as).

CONVOLUTED: coiled or twisted, like folds on the surface of the brain.

CORTINA: weblike strands extending from the stipe to the margin of the pileus.

CRENULATE: scalloped.

CRISTATE: with small crests or ridges.

CUTICLE: the thin outer layer of the pileus.

CYSTIDIA: sterile cells on the hymenium of a Basidiomycete.

DECURRENT: extending down the stipe.

DELIQUESCE: to turn to liquid.

DEPRESSED: having the central part sunken below the margin.

DEXTRINOID: a reddish-brown reaction (of spores) in Melzer's solution.

DISC: the central part of the pileus.

DISCOID: disc-shaped.

ECCENTRIC: having the stem attached on or toward one side.

ECHINULATE: spiny.

ELLIPSOID: elongate and round at both ends, with sides convex.

EMARGINATE: notched on the margin.

ENDOPERIDIUM: the inner layer of the covering of a puffball.

EPIGEOUS: growing on or above the ground.

EPIPHRAGM: a lid covering the cup in bird's-nest fungi.

EQUAL: of even diameter (describing the stipe).

EVANESCENT: soon disappearing.

EXOPERIDIUM: the outermost covering of a puffball.

FARINACEOUS: having the odor of grain.

FIBRIL: a small thread.

FIBRILLOSE: having minute fibers.

FILAMENTOUS: threadlike.

FILIFORM: threadlike.

FIMBRIATE: fringed.

FLOCCOSE: cottony or woolly.

FREE: describing gills not attached to the stipe.

FRIABLE: easily broken or crumbled.

FUNICULUS: a thin cord attaching the peridiole to the inner surface of the cup in some bird's-nest fungi.

FUSIFORM: tapering at both ends.

GERM PORE: a minute aperture on the end of a spore.

GLABROUS: smooth.

GLANDULAR: having moist or sticky dots on the surface.

GLEBA: the spore mass within a puffball.

GLOBOSE: spherical.

GLUTINOUS: covered with a gelatinous layer.

GREGARIOUS: growing in groups.

GUTTULATE: containing one or more oil drops.

HALLUCINOGENIC: producing hallucinations.

HOST: the organism, dead or alive, on which a fungus lives.

HYALINE: colorless.

HYGROPHANOUS: becoming watery when moist.

HYGROSCOPIC: capable of absorbing moisture.

HYMENIUM: the spore-bearing surface of a fleshy fungus.

HYPHAE: threadlike forms; the filamentous, branching aggregation of cells which compose a mycelium.
HYPOGEOUS: fruiting underground.

IMBRICATE: overlapping like shingles.
INFUNDIBULIFORM: funnel-shaped.
INROLLED: rolled down and inward.
INTERVENOSE: having cross veins.
INVOLUTE: rolled inward.

KOH: potassium hydroxide.

LACUNOSE: markedly pitted or indented.
LAMELLAE: gills.
LAMELLULAE: small alternating gills which do not extend all the way from the margin to the center.
LANCEOLATE: spear-shaped.
LATEX: fluid present in some mushrooms.
LIGNICOLOUS: growing on wood.

MEMBRANOUS: thin and pliant like a membrane.
MICRON: one-thousandth of a millimeter; symbol μ. (*mu*).
MYCELIUM: a mass of hyphae comprising the vegetative part of a fungus plant.
MYCOLOGICAL: pertaining to fungi.
MYCORRHIZA: the symbiotic relationship between a fungus and the roots of another plant.

NUCLEATE: containing oil-like globules.

OBTUSE: blunt.
OCHRACEOUS: yellowish.
OLIVACEOUS: olive-color.
ORNAMENTED: referring to spines, warts, or ridges on spores.
OVOID: egg-shaped.

PARAPHYSES: sterile filaments.

PARASITISM: subsisting on other living organisms.

PEDICEL: a short, stemlike process attached to the spores of some puffballs.

PELLICLE: the skin-like layer on the surface of the pileus.

PERIDIOLE: a small, round body containing spores in bird's-nest fungi.

PERIDIUM: the wall or covering of a Gasteromycete.

PERITHECIUM: a small, flask-shaped structure in which asci may be borne.

PILEATE: having a distinct cap or pileus.

PILEUS: the cap of the fruiting body.

PLANE: flat.

PLEUROCYSTIDIA: sterile cells on the sides of a gill.

PLICATE: pleated or folded.

POLYPORE: one of the family Polyporaceae.

POROID: resembling pores.

PRUINOSE: with a bloom or a powdery appearance.

PSEUDORHIZA: a false root.

PSYCHOTROPIC: acting on the mind.

PUBESCENT: having minute hairs.

PYRIFORM: pear-shaped.

PYXIDATE: cup-shaped on top.

RECEPTACULUM: the stalk of a stinkhorn.

REFLEXED: bent backward.

RENIFORM: kidney-shaped.

REPAND: having the margin turned up.

RESUPINATE: with the upper surface of the pileus resting on the substratum and the gills facing upward or outward.

RETICULATE: having netlike ridges.

RHIZOMORPH: a rootlike cord of compacted mycelium at the base of the stipe.

RUGOSE: roughened or wrinkled.

SAPROPHYTISM: subsisting on dead organic material.

SCROBICULATE: marked with shallow depressions.

SCROBICULATION: a pit

SCURFY: rough.

SEPTATE: having partitions.

SEPTUM: a partition.

SERRATE: toothed or with sawlike edge.

SESSILE: lacking a stipe.

SETAE: minute bristles.

SINUATE: describing gills notched next to the stipe.

SPATHULATE: spoon-shaped.

SPHAEROCYST: large round cells in the Russulaceae.

SPICULES: small spines.

SPOROPHORE: the fruiting body of a mushroom, its spore-producing organ.

SQUAMULES: small scales.

SQUAMULOSE: covered with small scales.

SQUARROSE: covered with erect or recurved, pointed scales.

STELLATE: star-shaped.

STERIGMA (pl. STERIGMATA): a spine bearing a spore on the tip of a basidium.

STIPE: the stem of the fruiting body.

STIPITATE: having a stipe (stem).

STRIATE: with radiating lines.

SUB- : a prefix meaning almost or nearly.

SULCATE: grooved.

SYMBIOSIS: a state of mutual interdependence between two different organisms.

TERRESTRIAL: growing on the ground.

TOMENTOSE: minutely hairy.

TOMENTUM: matted, woolly hairs on the surface.

TRAMA: the fleshy portion of the pileus or of the gill tissue, between the outer layers of the hymenium.

TRUNCATE: squared off.

TUBERCULATE: having small warts.

UMBILICATE: having the center of the pileus depressed.
UMBO: a raised area in the center of the pileus.
UMBONATE: having an umbo.
UNEQUAL: describing the condition in which some gills are shorter than others and do not reach the stipe.
UNGULATE: hoof-shaped.

VENTRICOSE: enlarged in the middle.
VINACEOUS: wine-color.
VIRGATE: streaked with fibrils, usually of a different color.
VISCID: sticky to the touch.
VOLVA: a cuplike sheath at the base of the stipe.

ZONATE: marked with concentric bands of color.

LIST OF ABBREVIATED NAMES
OF DESCRIBERS OF FUNGI

Afz.	A. Afzelius
Alb.	J. B. von Albertini
Atk.	F. Atkinson
Berk.	J. M. Berkeley
Bolt.	J. Bolton
Boud.	E. Boudier
Bres.	A. J. Bresadola
Br.	C. E. Broome
Bull.	P. Bulliard
Cke.	M. C. Cooke
Curt.	M. A. Curtis
D.C.	A. DeCandolle
Dicks.	J. Dickson
Fr.	E. M. Fries
Fckl.	K. W. G. L. Fuckel
Gill.	C. C. Gillet
Hark.	H. W. Harkness
Henn.	P. Hennings
Hook.	W. J. Hooker
Jacq.	N. Jacquin
Jung.	F. F. Junghuhn
Kalch.	K. Kalchbremer
Karst.	P. Karsten
Kauf.	C. Kauffman
Kl.	J. F. Klotzsch
Krombh.	J. V. von Krombholz
Kühn.	R. Kühner
Kumm.	P. Kummer
Lamb.	J. B. E. Lambotte
L.	C. von Linnaeus

Lund.	S. Lundell
Mass.	G. Massee
Morg.	A. P. Morgan
Müll.	O. F. Müller
Murr.	W. A. Murrill
Nannf.	J. A. Nannfeldt
Nyl.	W. Nylander
Opat.	W. Opatowski
Pat.	N. Patouillard
Pers.	C. H. Persoon
Phill.	W. Phillips
Pk.	C. H. Peck
Pom.	R. Pomerleau
Quél.	L. Quélet
Raitv.	A. Raitviir
Rick.	A. Ricken
Romag.	H. Romagnesi
Sacc.	P. A. Saccardo
Schaeff.	J. C. Schaeffer
Schroet.	J. Schroeter
Schw.	L. D. Schweinitz
Scop.	J. A. Scopoli
Secr.	L. Secretan
Sing.	R. Singer
Sow.	J. Sowerby
Trott.	A. Trotter
Tul.	E. L. Tulasne
Underw.	L. M. Underwood
Vitt.	C. Vittadini
Wallr.	C. F. Wallroth
Weinm.	J. A. Weinmann

WESTERN MYCOLOGICAL COLLECTIONS AND SOCIETIES

Herbarium collections of dried fungi may be found in several of our universities, colleges, and museums, and most have specialists on the staff who will assist in identification of fresh specimens. A list of some of these institutions follows:

California Academy of Sciences
Denver Botanic Gardens
Humboldt State University
San Francisco State University
Santa Barbara Museum of Natural History
University of California (Berkeley and Los Angeles)
University of Idaho
University of Washington

There are many regional and local mycological societies whose membership is composed primarily of amateurs interested in learning more about wild mushrooms. These organizations have regular meetings and lectures on interesting mycological subjects. During the mushroom season members often bring in fresh specimens, which are labelled and placed on display. Most societies also have a number of field trips or forays on weekends during the fruiting period. Such labeled displays and field experiences with competent leaders provide an excellent means of becoming acquainted with your local fungi. Some of the societies also have an annual Fungus Fair, when several hundred freshly gathered species of identified fungi may be on exhibit. There may be special displays showing poisonous species as well as choice

274

table delicacies. Some of these regional mushroom societies as well as a few national organizations are listed below:

Colorado Mycological Society
909 York Street
Denver, Colorado 80206

Hoquiam Mushroom Club
Rt. 2, Box 193
Hoquiam, Washington 98550

Kitsap Peninsula Mushroom Society
1132-A Magnuson Way
Bremerton, Washington 98310

Lincoln County Mycological Society
P.O. Box 94
Siletz, Oregon 97380

Los Angeles Mycological Society
1615 N. Beverly Glen Blvd.
Los Angeles, California 90024

Mycological Society of San Francisco
P.O. Box 904
San Francisco, California 94101

Mycological Society of Santa Barbara
3194 Via Real
Carpinteria, California 93013

North Idaho Mycological Association
Rt. 2, Box 186
Post Falls, Idaho 83854

South Idaho Mycological Association
719 Eighth Ave.
S. Nampa, Idaho 83651

Oregon Mycological Society
6548 S.E. 30th Avenue
Portland, Oregon 97202

Puget Sound Mycological Society
200 2nd Avenue N.
Seattle, Washington 98109

Snohomish Mycological Society
12225 13th Avenue S.E.
Everett, Washington 99203

Spokane Mushroom Club
443 W. 25th Ave., North
Spokane, Washington 99203

Tacoma Mushroom Society
1505 S. Mason
Tacoma, Washington 98405

Tri-Cities Mycological Society
1628 W. Clark
Pasco, Washington 99301

Twin Harbors Mushroom Club
Rt. 2, Box 193
Hoquiam, Washington 98550

REFERENCES

Bandoni, R. J. and A. F. Szczawinski, 1964. Guide to Common Mushrooms of British Columbia. Brit. Col. Provincial Museum, Handbook No. 24.

Bigelow, H. E., O. K. Miller, and H. D. Thiers. 1976. A New Species of *Omphalotus*. Mycotaxon 3 (3): 363-372.

Corner, E. J. H. 1950. A Monograph of *Clavaria* and Allied Genera. Annals of Botany Memoirs, no. 1. London: Oxford University Press.

Dissing, H. 1966. The Genus *Helvella* in Europe. Dansk Botanisk Arkiv 25 (1): 1-172.

Doty, M. S. 1944. *Clavaria*, the Species Known from Oregon and the Pacific Northwest. Oregon State Monographs. Studies in Botany, no. 7. Corvallis: Oregon State University.

Duffy, T. J., and P. P. Vergeer. 1977. California Toxic Fungi. Monograph no. 1. Mycological Society of San Francisco.

Gilkey, H. M. 1939. Tuberales of North America. Oregon State Monographs. Studies in Botany, no. 1. Corvallis: Oregon State University.

Hall, D., and D. E. Stuntz. 1972. Pileate Hydnaceae of the Puget Sound Area III; Brown-Spored Genus *Hydnellum*. Mycologia 44: 560-590.

Harkness, H. W. 1899. California Hypogaeous Fungi. Proceedings California Academy of Sciences, 3rd ser., Botany, 1: 241-292.

Harrison, K. A. 1961. The Stipitate Hydnums of Nova Scotia. Publication 1099. Ottawa: Canadian Department of Agriculture.

Hesler, L. R., and A. H. Smith. 1963. North American Species of *Hygrophorus*. Knoxville: University of Tennessee Press.

_____. 1965. North American Species of *Crepidotus*. New York: Hafner.

Kauffman, C. H. 1918. The Agaricaceae of Michigan. Biological Series, no. 26. Lansing: Michigan Geological and Biological Survey.

Kempton, P. E., and V. L. Wells. 1972. Studies on the Fleshy Fungi of Alaska IV; a Preliminary Account of the Genus *Helvella*. Mycologia 42: 940-959.

Largent, D. L. 1977. The Genus *Leptonia*. Bibliotheca Mycologica 55. Lehre, Germany: J. Cramer.

_____. 1973. How To Identify Mushrooms. Vol. I: Using Only Macroscopic Features. Eureka, Calif.: Mad River Press.

_____, and D. Johnson. 1977. How To Identify Mushrooms. Vol. III: Microscopic Features. Eureka, Calif.: Mad River Press.

_____, and H. D. Thiers. 1977. How To Identify Mushrooms. Vol. II: Field Identification of Genera. Eureka, Calif.: Mad River Press.

Lincoff, G., and D. H. Mitchel. 1977. Toxic and Hallucinogenic Mushroom Poisoning. New York: Van Nostrand.

Mains, E. B. 1956. North American Species of the Geoglossaceae; Tribe Cudonieae. Mycologia 48: 694-710.

Marr, D. C., and D. E. Stuntz. 1973. *Ramaria* of Western Washington. Bibliotheca Mycologia 38. Lehre, Germany: J. Cramer.

McKnight, K. H. 1969. A Note on *Discina*. Mycologia 61: 614-630.

Menser, G. P. 1977. Hallucinogenic and Poisonous Mushroom Field Guide. Berkeley, Calif.: And/Or Press.

Miller, O. K. 1964. Monograph of *Chroogomphus* (Gomphidiaceae). Mycologia 56: 526-549.

_____. 1968. Interesting Fungi of the St. Elias Mountains, Yukon Territory, and Adjacent Alaska. Mycologia 60: 1190-1203.

_____. 1969. Notes on Homobasidiomycetes from Northern Canada and Alaska. Mycologia 61: 840-844.

_____. 1969. Notes on Gastromycetes of the Yukon Territory and Adjacent Alaska. Can. J. Botany 47: 247-250.

_____. 1971. The Genus *Gomphidius* Fries, with a Revised Description of the Gomphidiaceae and Keys to the Genera. Mycologia 63: 1129-1163.

_____. 1972. Mushrooms of North America. New York: Dutton.

_____, and D. F. Farr. 1975. An Index of the Common Fungi of North America (Synonyms and Common Names). Bibliotheca Mycologia 44. Lehre, Germany: J. Cramer.

Orr, R. T., and D. B. Orr. 1962. Mushrooms and Other Common Fungi of the San Francisco Bay Region. Berkeley: University of California Press.

_____. 1968. Mushrooms and Other Common Fungi of Southern California. Berkeley: University of California Press.

Overholts, L. O. 1953. The Polyporaceae of the United States, Alaska, and Canada. Ann Arbor: University of Michigan Press.

Petersen, R. H. 1975. *Ramaria* Subgenus *Lentoramaria*, with Emphasis on North American Taxa. Bibliotheca Mycologia 43. Lehre, Germany: J. Cramer.

Schaffer, R. L. 1962. The Subsection *Compactae* of *Russula*. Brittonia 14: 254-284.

_____. 1968. Keys to Genera of Higher Fungi, 2nd ed. Ann Arbor: University of Michigan Biological Station.

Seaver, F. J. 1942. The North American Cup-Fungi (Operculates). New York: Seaver.

_____. 1951. The North American Cup-Fungi (Inoperculates). New York: Seaver.

Singer, R. 1962. The Agaricales in Modern Taxonomy. Lehre, Germany: J. Cramer.

Smith, A. H. 1947. North American Species of *Mycena*. Ann Arbor: University of Michigan Press.

_____. 1968. The Cantharellaceae of Michigan. Michigan Botanist 7: 143-167.

_____. 1975. A Field Guide to Western Mushrooms. Ann Arbor: University of Michigan Press.

_____, and L. R. Hesler. 1968. North American Species of *Pholiota*. New York: Hafner.

_____, and R. Singer. 1964. A Monograph of the Genus *Galerina*. New York: Hafner.

_____, and H. V. Smith. 1973. The Non-Gilled Fleshy Fungi. Dubuque, Iowa: Wm. C. Brown Co.

_____, and H. D. Thiers. 1964. A Contribution Toward a Monograph of North American Species of *Suillus*. Ann Arbor, Mich.

_____, H. D. Thiers, and O. K. Miller. 1965. The Species of *Suillus* and *Fuscoboletinus* of the Priest River Experimental Forest and Vicinity, Priest River, Idaho. Lloydia 28: 120-138.

_____, H. D. Thiers, and R. Watling. 1966. A Preliminary Account of the North American Species of *Leccinum*, Section *Leccinum*. Michigan Botanist 5: 131-179.

_____, H. D. Thiers, and R. Watling. 1967. A Preliminary Account of the North American Species of *Leccinum*, sections *Luteoscabra* and *Scabra*. Michigan Botanist 6: 107-154.

Stuntz, D. D. 1977. How To Identify Mushrooms. Vol. IV: Keys to Families and Genera. Eureka, Calif.: Mad River Press.

Thiers, H. D. 1965. California Boletes I. Mycologia 57: 524-534.

_____. 1966. California Boletes II. Mycologia 58: 815-826.

Tylutki, E. E. 1962. Some Edible and Poisonous Mushrooms of Idaho. J. Acad. Sci., Special Research Issue 1.

Weber, N. S. 1972. The genus *Helvella* in Michigan. Michigan Botanist 11: 147-201.

Wells, V. L., and P. E. Kempton. 1968. A Preliminary Study of *Clavariadelphus* in North America. Michigan Botanist 7: 35-57.

Zeller, S. M., and A. H. Smith. 1964. The Genus *Calvatia* in North America. Lloydia 27: 148-186.

INDEX